CHENG TAINING 程泰宁
Architecture Works 2009-2014　建筑作品选 2009-2014

中国建筑工业出版社

程泰宁

中国工程院 院士
中国建筑设计大师
教授,博士生导师
东南大学建筑设计与理论研究中心主任
中国联合工程公司总建筑师
筑境设计主持人
梁思成建筑奖获得者

CHENG TAINING

Academician of Chinese Academy of Engineering

China Architecture Design Master

Professor, Doctorial Supervisor

Director of Architectural Design & Theory Research Centre of Southeast University

Chief Architect of China United Engineering Corporation

Chairman,CCTN Design

Winner of Liang Sicheng Architectural Award

立足此时　　立足此地　　立足自己
BASED ON NOW　　BASED ON HERE　　BASED ON MYSELF

CONTENTS 目录

PREFACE	6	代序
DESIGN WORKS	15	设计作品选
Nanjing Museum (Phase Ⅱ)	16	南京博物院改扩建工程
Longquan Celadon Museum	52	龙泉青瓷博物馆
China Museum of Sea Salts	70	中国海盐博物馆
Ningxia Theater	86	宁夏大剧院
China Ports Museum	112	中国港口博物馆
Xiangtan Urban Planning Exhibition Hall and Museum	122	湘潭城市规划展览馆及博物馆
Nanxun Administrative Center, Huzhou	136	湖州南浔行政中心
Zhaoshan Two Type Industrial Developing Center, Hunan	146	湖南昭山两型发展中心
Hangzhou Metropolis Xinyu Residential Area	164	杭州城市芯宇住宅小区
Hangzhou Qianjiang Financial Area	175	杭州钱江金融城概念方案
Conceptual Design of Daming Palace Site Museum, Xi'an	190	西安大明宫遗址博物馆概念方案
Wenling Museum	200	温岭博物馆
Suzhou Yue City Site Museum	210	苏州越城遗址博物馆
Conceptual Design of Jinyang New Town Exhibition Hall, Taiyuan, Shanxi	218	山西太原晋阳新城展示馆概念方案
Su Buqing Memorial Hall	222	苏步青纪念馆
Cangqian Campus (B Block) of Hangzhou Normal University	230	杭州师范大学仓前校区B组团
Yuehai Bay Hotel, Xiamen	236	厦门悦海湾酒店
Shanghai Shanshan Group Headquarters Building	248	上海杉杉控股总部大楼
Conceptual Design of Xidong New Town Cultural Center, Wuxi	254	无锡锡东新城文化中心概念方案
Proposal of Zhanjiang Cultural Center	262	湛江文化艺术中心方案
CHRONOLOGICAL LIST OF PROJECTS (2009-2014)	282	作品年表（2009-2014）
POSTSCRIPT	287	后记

代序

语言·意境·境界
——中国智慧在建筑创作中的运用

一

改革开放30年来，中国经济建设的成就有目共睹，但中国建筑的现状，似乎与这一发展进程不相匹配，"千城一面"和"缺乏中国特色"的公众评价，突显了我们所面临的困境。产生这一问题的原因是多方面的，但是应该看到，在建筑创作中，缺乏独立的价值判断和自己的哲学、美学思考，是其中一个十分重要的原因。

二

近百年来，中国现代建筑一直处在西方建筑文化的强势影响之下。从好处说，西方现代建筑的引入，推动了中国建筑的发展；从负面来讲，我们的建筑理念一直为西方所裹挟，在跨文化对话中"失语"，是一个不争的客观事实。虽然在这个过程中有不少学者、建筑师以至政府官员，在反思的基础上，倡导过"民族形式"、"中国风格"等等，但由于缺乏有力的理论体系作支撑，只是以形式语言反形式语言，以民粹主义反外来文化，其结果，只能停留在表面上而最后无疾而终。因此，建构自己哲学和美学思想体系、以支撑中国现代建筑的发展，是一个值得我们重视并加以研究的重要问题。

那如何来建构这样一个理论体系？我同意这样的观点，"中国文化更新的希望，就在于深入理解西方思想的来龙去脉，并在此基础上重新理解自己"〔乐黛云，（法）阿兰·李比雄〕据此，我们需要首先来了解一下西方现当代建筑的哲学和美学背景。

三

在西方，"20世纪是语言哲学的天下"。海德格尔说"语言是存在之家"，德里达说"文本之外无他物"，卡尔纳普则干脆把哲学归结为句法研究、语义分析。特别是近十几年"数字语言"的出现，似乎更加确立了"语言哲学"在西方的"统领地位"（以上参见李泽厚：《能不能让哲学走出语言》）。了解了西方这样的哲学背景，我们会很自然地想到，西方现当代建筑是不是在一定程度上也是"语言"的天下？耳熟能详的像"符号"、"原型"、"模式语言"、"空间句法"、"形式建构"，以至最新的"参数化语言"、"非线性语言"等。事实上，这些建筑"语言"都可以看作是西方语言哲学的滥觞。通过学术交流，这些"语言"也已经成了很多中国建筑师在创作中最常用到的词语。

对于这种现象如何看？

应该看到，"语言"包含着语义，特别是它对"只可意会不可言传"的建筑创作机制进行了理性的分析解读，值得我们借鉴。但同样应该看到，由于它在不同程度上忽视了人们的文化心理和情感，忽视了万事万物之间存在的深层次联系，很难完整地解释和反映建筑创作实际，因而这些"语言"常

古巴吉隆坡胜利纪念碑国际设计竞赛参赛方案 \ International Competition for Monument of Pig-Bay,Cuba

杭州黄龙饭店 \ Dragan Hotel,Hangzhou

加纳国家剧院 \ Ghana National Theatre,Accra

常是在流行一段时间以后光环渐失，在创作实践中并未起到"圣经"作用。

特别值得注意的是，以"语言"为本体，极易走入偏重"外象"的"形式主义"的歧路。我们已经明显地看到，从20世纪后半期开始，以"语言"为本体的哲学认知与后工业社会文明相结合，西方文化出现了一种从追求"本原"，逐步转而追求"图象化"、"奇观化"的倾向。法国学者盖德堡认为，西方开始进入一个"奇观的社会"；一个"外观"优于"存在"，"看起来"优于"是什么"的社会。在这种社会背景下，反理性思潮盛行，有的艺术家认为"艺术的本质在于新奇"，"只有作品的形式能引起人们的惊奇，艺术才有生命力"。他们完全否定传统、认为"破坏性即创造性、现代性"。了解了这样的哲学和美学背景就不难理解，一些西方先锋建筑师的设计观念和作品风格来自何处。对中国建筑师来说，我们在"欣赏"这些作品的时候是否也需要思考：这种以"语言"为哲学本体，注重外在形式，强调"视觉刺激"的西方建筑理念是否也有它的局限？我们能否走出"语言"，在建筑理论体系的建构上另辟蹊径？

四

实际上，百年来，一代代中国学者一直在进行中国哲学和美学体系的研究和探索。例如从王国维先生开始，很多学者就提出把"意境"作为一种美学范畴，试图建构一种具有东方特色的美学体系；近年来，著名学者李泽厚先生更是以"该中国哲学登场了"为主旨，提出了以"情本体"取代西方以"语言"为本体的哲学命题……。这些哲学和美学思考，是中国学者长时期来对东西方文化进行深入比较和研究的成果。尽管由于建筑的双重性，我们不能把建筑与文艺等同起来，但毫无疑问，这一系列研究对于我们建构当代中国建筑理论有重要的启迪。

从这些研究出发，结合中国建筑创作的现状和发展，我考虑，相对于西方以分析为基础、以"语言"为本体的建筑理念，我们可否建构以"语言"为手段、以"意境"为美学特征，以"境界"为本体这一具有东方智慧的建筑理念，作为我们在建筑上求变创新的哲学和美学支撑？我认为，这不仅是可能的，而且是符合世界建筑文化多元化发展需要的。

五

结合创作实践，我把建筑创作由表及里分解为三个层面：即：形（形式、语言）、意（意境、意义）、理（哲理、"境界"）。

六

第一个层面为"形"，即语言、形式。相对于西方对于"语言"的认知，中国传统文化的"大美

代序

马里会议大厦 \ Conference Building Mali,Bamako

杭州铁路新客站 \ Hangzhou New Railway Station

杭州国际假日酒店 \ Holiday Inn,Hangzhou

不言"、"天何言哉",禅宗的不立文字、讲求"顿悟",几乎抹杀了语言和形式存在的意义,这显然有些绝对化。而顾恺之的"以形写神"、王昌龄的"言以表意",则比较恰当地表达了语言形式和"意"、"神"的辩证关系。按此理解,语言只是传神表意的一种手段,而非本体。既为手段,那么,在创作中,建筑师为了更好地表达自己的设计理念,可选择的手段应该是多种多样的。特别是在建筑创作的三个层面中,较之"意"、"理"的相对稳定,"语言"会随着时代的发展而不断变化,建筑师需要在充分掌握中外古今建筑语言的基础上,不断地转换创新。我以为,走出西方建筑"语言"的藩篱,摆脱"语言"同质化、程式化的桎梏,我们在语言创新方面将会有更为广阔的视界,在重新审视中国传统文化中"大气中和"、"含蓄典雅"等语言特色的同时,在建筑形式美、语言美的探索上力争有自己的新的突破。

七

建筑创作第二个层面为"意",即意境,意义。这里我们重点谈"意境"。

上面我们曾提到中国传统文化否定"语言"的绝对化倾向,但我们更要看到"大美不言""大象无形"的哲学思辨,也赋予了中国传统绘画、文学包括建筑以特有的美学观念。从很多优秀的传统建筑中可以看出,人们已超越"语言"层面,通过空间营造等手段,进而探索意境、氛围和内心体验的表达,把人们的审美活动由视觉经验的层次引入静心观照的领域,追求一种言以表意、形以寄理、情境交融、情溢象外的审美境界。这给建筑带来了比形式语言更为丰富,也更为持久的艺术感染力。

"意境"、"情境合一",是一种有很高品位的中国式的审美理想,是建构有中国特色美学体系的基础。对"意境"的理解和塑造,是中国建筑师与生俱来的文化优势,不少建筑师已经进行了有益的探索,我想,进一步自觉地开展这方面的研究和探索,对于我们摆脱"语言"本体的束缚,在理论和实践上实现突破创新,是十分重要的。

八

建筑创作的第三个层面为"理"、哲理。我认为,建筑创作的哲理——亦即"最高智慧",是"境界"。

何谓"境界"?王国维在《人间词话》的手稿中说,"不期工而自工"是文艺创作的理想境界;有学者进一步解释说,"妙手造文,能使其纷沓之情思,为极自然之表现"即为"境界"(周编《人间词话》P001、P002)。结合建筑创作,我认为这里包含着两方面的含义:

其一,从"天人合一"、万物归于"道"的哲学认知出发,要看到,身处大千世界,建筑从来不是一个孤立的单体,而是"万事万物"的一个组成分子。在创作中,摆正建筑的位置,特别注意把建

宁波高教园图书信息中心
\ Ningbo Higher Education Park Books Information Center

联合国国际小水电中心
\ United Nations International Hydraulic Power Center

上海公安局办公指挥大楼
\ Commanding Center of Shanghai Pubilc Security Bureau

筑放在包括物质环境和精神环境这样一个大环境、大背景下进行考量，既重分析、更重综合，追求自然和谐；既讲个体、更重整体，追求有机统一；使建筑、人与环境呈现一种"不期工而自工"的整体契合、浑然天成的状态，是我们所追求的"天人境界"，也是我们所需要建构的建筑观与认识论。

其二，"境界"不仅诠释并强调了建筑和外部世界的内在联系，而且还揭示了建筑创作本身的内在机制。以"境界"为本体，我们可以看到，在建筑创作中，功能、形式、建构，以至意义、意象等等理性与非理性因素之间，并不遵循"内容决定形式"或"形式包容功能"这类线性的逻辑思维模式，也很难区分哪些是"基本范畴"和"派生范畴"（美·戴维·史密斯·卡彭《建筑理论》）。在创作实践中，建筑师所建构的，应该是一个以多种因素为节点的、相互联结的网络。当我们游走在这个网络之中，不同的建筑师可以根据自己理解和创意，选择不同的切入点，如果选择的切入点恰当，我们的作品不但能够解决某一个节点（如形式）的问题，而且能够激活整个网络，使所有其他各种问题和要求相应地得到满足。这种使"纷沓的情思"得到"极自然表现"的"自然生成"，是我们追求的创作"境界"。因此，从语言哲学和线性逻辑思维模式中解放出来，以"境界"这一具有中国智慧的哲学思辨来诠释建筑创作机制，建构一种符合建筑创作内在规律的"理象合一"的方法论，将使建筑创作的魅力和价值能够更加充分地显示出来。

此外，以境界为本体，还可以使我们更好地理解并运用那些充满东方智慧的、具有创造性的思维方式。例如直觉、通感、体悟……。这些具有创造性的思维活动（方式），需要在反复实践和思考中获得，它也体现了一种建筑境界。

以上，很简单地谈了我对"语言"、"意境"与"境界"的理解，以下，我想结合我的一些作品，进一步谈谈我对建筑创作中"语言"（形式）、"意境"（空间）和"境界"（环境）这三个层面的思考以及在创作中的具体运用。

（2014年11月在"第十届亚洲建筑国际交流会"的主旨讲演）

2014年10月

PREFACE

APPLICATION OF EASTERN WISDOM IN THE ARCHITECTURAL DESIGN CREATION
LANGUAGE, CONCEPTION AND "JINGJIE" ①

李叔同（弘一大师）纪念馆
\ Li Shutong (Hong Yi Master) Memorial Hall

绍兴鲁迅纪念馆 \ Shaoxing Lu Xun Memorial Hall

浙江美术馆 \ Zhejiang Art Museum

1.

During the thirty years since reform and opening up, China has been integrated into the world at an incredible fast speed. However, the current situation of Chinese architectural creation is unsatisfactory. Cities with similar look and lacking of Chinese cultural characteristics are widely questioned and criticized by the public. Reasons to these problems are various, but we must acknowledge that, the lack of independent value judgment, philosophy system and aesthetic thoughts, are important reasons for the current realistic predicament of Chinese architecture.

2.

For hundreds of years, modern Chinese architecture has been under a strong influence of western architectural culture. In a good way, the introduction of western contemporary architecture has promoted the Chinese architectural development; while in the other way, we are unwittingly accepting the strong influence of Western culture due to the breakage of Chinese traditional culture and the "deconstructed" value systems. Although during this period of time, many scholars and architects and even government officials have advocated the "national style" and the "Chinese-style" on the base of reflection, since there is no in-depth and systematic theoretical thinking as support, it appears more often to use one kind of form against another, and to use populism against foreign culture. The results can only stay on the surface. Therefore, constructing the philosophical and aesthetic system reflecting Chinese wisdom and having universal value, to support and promote the development of modern Chinese architecture, is a subject worthy of lots of attention and research. How to construct such a theoretical system? I would agree with this strategy that "the hope of Chinese culture renewal, lies deeply in a complete understanding of the western thinking, and a re-understanding of our own based on this."(Cross-culture Dialogue, *Yue Daiyun&Alain Le Pichon*) Thus, it is needed for us to take a look around the philosophical and aesthetical background of western modern and contemporary architecture.

3.

In western countries, it is believed that "the 20th century was dominated by the philosophy of language". Heidegger said that "language is the home of existence" (Heidegger, 1997). And Jacques Derrida pointed that "Il n'y a pas de hors-texte" (Outside text, there's nothing) while Rudolf Carnap generalized that philosophy was a syntactic research and a semantic analysis. Especially the emergence of "digital language" in recent years, has established the leading position of "language philosophy" in western countries. With the understanding of such western philosophy background, we would naturally come to mind that if the western contemporary architecture is, to some extent, also the world of "language". Phrases such as "semiotic language", the "pattern language" and the latest "parameterized language" impress us more and more nowadays. In fact, these architectural "languages" can be seen as the expressions of western language philosophy. And through academic communication, these "languages" have already been used by many Chinese architects in their creations.

How should we evaluate this phenomenon?

First of all, we must see and learn from it that "languages" contain some kind of semantic meaning and play a certain role in expressing the architectural creation mechanisms which is always known as something could hardly been expressed. But we should also notice that, they neglect the human's cultural psychology and emotion, and ignore the various and initial relationship existing among people, nature environment and all matters including architectures, which makes it difficult to interpret and reflect the reality of architectural creation completely. Thus these "languages" always gradually fade out after a period of time, instead of acting as the general principle in design and creation.

And another remarkable fact is that the "language" ontology easily leads to formalism which stresses on "out looking". We could obviously perceive that from the second half of 20th century, the philosophical cognition with the "language" as ontology has led in the pursuit tendency from "origin" to "image", and even "spectacle". The French scholar Guy Debord argued that western countries had entered a "society of spectacle", in which "appearance" is superior to "existence", and "how it look" is superior to "what it is". In such social context, the anti-rational thought prevails in culture and arts. Some artists believed that "the essence of art lies in novelty", and "only the form of work can arouse people's curiosity, and gives vitality to art". They completely denied tradition, and argued that "destructivity equals to creativity and modernity." With such understanding of philosophical and aesthetic backgrounds, it is not difficult to comprehend that where do those design concepts and styles of the western avant-garde architects come from. For Chinese architects, we should think about while "appreciating" these works: whether such architectural concept (features "language" as ontology, concerns on the external form, and emphasizes on the "visual stimulation") also has its limitations or not? And whether we can go beyond the "language" and find other ways to construct the architectural theory system?

4.
As a matter of fact, during the past century, generations of Chinese scholars have devoted themselves to the exploration and research of Chinese philosophy and aesthetics. For example, starting from Mr. Wang Kuo-wei, many scholars had proposed to consider "artistic conception" as an aesthetic category, and attempted to construct an aesthetic system with eastern characteristics. In recent years, the famous scholar Mr. Li Zehou even use the "ontology of emotion" to replace the western ontology on "language", under the theme "It is time for Chinese philosophy". (Li, 2011) These philosophical and aesthetic thinking is achievements made from precise research and in-depth comparison of eastern and western cultures, which is done by generations of Chinese scholars over a long time. Although we cannot equate architecture with art due to dual nature of architecture. Undoubtedly, this series of studies are important enlightenment for the construction of contemporary Chinese architectural theory.

Proceeding from these studies, considering the current situation and development of architectural creation in China, in my viewpoint, instead of using the western architectural concept which regards analysis as basis and "language" as ontology, could we construct an architectural concept of eastern wisdom which regards "language" as approaches, "aesthetic conception" as aesthetic characteristics, and "Jingjie" as ontology, to be the philosophical and aesthetical support of our renovation in architectural creation? I think it is not only possible, but also in line with the tendency of the world's architectural culture diversity development.

5.
Integrated with creation practice, my thinking of the creation procedure can be divided into three levels from outward appearance to inner essence, namely: the "Xing"(form, language), the "Yi" (meaning, the artistic conception), and the "Li" ("Jingjie", the Chinese philosophy, refers to 'consummate realm').

PREFACE

四川建川博物馆·不屈战俘馆 \ Sichuan Jianchuan Museum·Undefeatable POW Museum

浙江宾馆商务别墅 \ Zhejiang Hotel Business Villa

加纳国家剧院 \ Ghana National Theater

6.

The first level is "xing" (means shape in Chinese), referring to "language" and form. In comparison to the western cognition for "language", in Chinese traditional culture, the "greatest beauty is beyond words", in Zen Buddhism, the emphasis on "sudden insight" and "communication of minds instead of written words", these ideas seem to be too absolute that they almost completely obliterate the meaning of language and form. However, the "conveying spirit through form" by Gu Kaizhi and "use words to express thinking" by Wang Changling, are more appropriate to explain the dialectic relationship between the language form and the "meaning" and "spirit". Based on such understanding, the language is just a way of conveying spirit and expressing meaning. Especially in the three levels of architectural creation, compared with "Yi" and "Li" who are relatively stable, the "language" could constantly change with the development of the times. So architects need to fully grasp the "languages" in modern or ancient times, in China or elsewhere, and on such basis, to make continuous innovation and transformation. I think, once we could cross the barriers of western architectural "language" and getting rid of the shackle of the "language" homogenization and stylization, we will have a broader outlook and make breakthrough in architectural form and "language".

7.

The second level is "Yi", referring to the artistic conception and the meaning. I would like to focus on "artistic conception".

It has been mentioned above that the traditional Chinese culture which denies the absolute tendency of "language", but we still have to admit the unique aesthetic concepts endowed by the philosophical thinking as the "greatest beauty is beyond words" and "greatest image is without form" to traditional Chinese painting, literature, and architecture. It can be seen from many other excellent traditional architecture that people have transcend the "language" level, exploring the expression of mood, atmosphere and inner experience through the form and space, and thus led aesthetic activities of people from visual experience to psychological experience. This gives the architecture more profound, richer, and long-lasting artistic appeal than the formal language. The aesthetic conception, is an eastern aesthetic ideal of a high level, and is the basis for constructing eastern aesthetic system. To understand and construct this "artistic conception", Chinese architects enjoy their cultural advantages in their blood. Many architects have already made beneficial explorations. In my opinion, the further conscious research and exploration in this area are very important for us to get rid of the binding of the "language" ontology and achieve breakthrough innovation in both theory and practice.

8.
The third level is "Li", referring the philosophy. In my opinion, the philosophy of architectural creation, which is also "the highest level", is "Jingjie"(translated as the 'consummate realm').

What is "Jingjie"? Wang Kuo-wei said in his manuscript of *Jen-ChienTz'u-hua: a Study in Chinese Literary Criticism* that "delicate words come out natually" is the highest level of artistic creation. Some scholars have further explained that "Jingjie" is when "words come from nature, express the nature and the feeling". Combined with architectural creation, I think "Jingjie" has two kinds of meanings:

Firstly, proceeding from the philosophical cognition of "Harmony between the Human and the Nature", we must see that in the whole world, architecture is not an isolated individual, but a part of all the beings. In architectural creation, we need a macro and overall mode of thinking which gives equal emphasis on analysis and integration; pursues harmony between the building and the nature; stresses on both individual and integral, and try to achieve organic unity. The pursuit of overall fit between architecture and environment (physical and mental), and the pursuit of "naturally expressing" after integrating multiple factors into architectural creation can reflect the wisdom of the architect. It is the architectural view and epistemology we need to construct.

Secondly, "Jingjie" not only explains the internal relationship between architecture and the outside world, but also explores the internal mechanism of architectural creation itself. When we take the "Jingjie" as the architectural ontology, we could see that elements of architectural creation, including function, form, construction, and even spiritual factors such as meaning, do not precisely follow the linear thinking of "what decides what", and it is hard to say which are basis and which are derivation(David Smith Capon, *Architectural Theory*). In architectural creation, what we need to construct is a network consisted of various elements. When we are wandering in this network, different architects could choose different element as an entry point to start design. If the entry point is well chosen, we can not only solve a unique problem of this element but also stimulate the whole network to meet all the other needs. This kind of creation which derives from nature and expressing (the nature and the feeling) naturally, is just the "Jingjie", the creation level, the "state of mind" we need to achieve. So, in order to fully present the value and beauty of architectural creation, we need to jump out of the "language" philosophy and the linear logic of thinking, to use "Jingjie", the Chinese philosophical state, to explain the architectural creation mechanism, and to construct a methodology which fits the inner principles of architectural creation.

Besides, ontology on "Jingjie", would help us better understand and apply the eastern way of thinking which is sapiential and creative, such as "Zhijue"(instinct), "Tonggan"(feeling), "Tiwu"(insight), etc. These creative ways of thinking could only be achieved after a large amount of thinking and practicing, which presents "Jingjie" again.

All above is my understanding of "language", "artistic conception" and "Jingjie". In following chapters,there are examples of application in architectural creation towards these three levels.

(Speech on *the 10th International Symposium on Architectural Interchanges in Asia*, November, 2014)

Cheng Taining
Oct.2014

(1) JINGJIE: Chinese vocabulary, means 'consummate realm' in Eastern philosophy.

设计作品选　DESIGN WORKS

南京博物院改扩建工程
NANJING MUSEUM (PHASE II)

合作者 王幼芬、王大鹏、柴 敬、张朋君、刘辉瑜、骆晓怡、应 瑛
合作单位 江苏省建筑设计院
设 计 2008 / 竣 工 2014

Co-designers: Wang Youfen,Wang Dapeng,Chai Jing,Zhang Pengjun,Liu Huiyu,
Luo Xiaoyi,Ying Yin
Cooperation Company: Jiangsu Provincial Architectural D&R institute LTD
Design Time: 2008 / Completion Time: 2014

补白·整合·新构
Blank Filling·Integration·New Construction

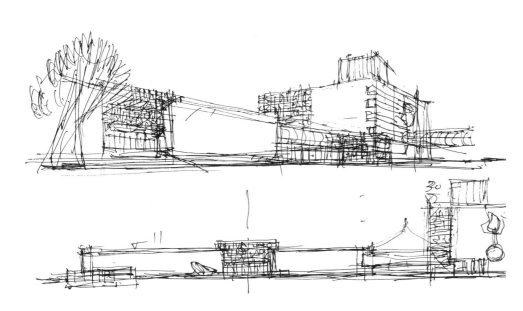

全景鸟瞰（陈 畅 摄影）\ Panoramic Aerial View (Photographer, Chen Chang)

南京博物院位于南京中山门内西北侧，其前身系蔡元培等人于1933年创建的国立中央博物院筹备处。南京博物院建筑由徐敬直先生中标设计，后由梁思成先生主持了方案修改工作。工程于1933年动工。因抗战爆发，当初规划的自然、人文、艺术三馆仅建成人文馆（历史馆），后在1999年新建了艺术馆。随着时代进步，原有展馆已无法适应现代博物馆展陈要求，因此南京博物院二期工程立项。

Nanjing Museum, located to the northwest of Zhongshan Gate, has its predecessor as National Central Museum Preparatory Office which was established in 1933 by Cai Yuanpei, Due to the breakout of Anti-Japanese War, only the Humanity Museum (History Museum) designed by Mr.Liang Sicheng was constructed of the three originally planned museums. The other two are the Nature Museum and the Art Museum. The latter was built in 1999. With the progression of age, the original exhibition museum could not fit the exhibition requirements of a modern museum. Consequently, the project of Nanjing Museum Phase II was approved.

改扩建方案的设计理念是：补白·整合·新构。

"补白"是对不同时期的建筑和场地环境进行分析梳理，将新扩建的建筑恰如其分地布置在合适的位置，从而使得新老建筑以及建筑与场地环境相和谐。

"整合"，分别通过对新老建筑功能布局、交通流线体系、新老馆内外部空间、新老建筑形式与材料以及展览与休闲功能五个方面进行整合梳理，使其达到一体化。

在"补白"与"整合"的基础上，设计重点通过对中轴空间、建筑形式、环境景观的整体塑造达到"新构"的目的。

"老大殿"原地抬升3m，既改善了原来建筑低于城市道路3m的不利现状，也减少了地下空间大面积的填挖土方，也为地面上下空间的流线组织创造了有利条件。

The proposal employs the design concept of "blank filling, integration and new construction".

"Blank filling" means to seek the harmoniousness between old and new buildings by analyzing and sorting the building and site environments of different periods, and by appropriately setting the new and expanded buildings in good places.

"Integration" means to achieve integration through the proper arrangement between the functional layout between old and new buildings, the traffic flow line system, the internal and external spaces of new and old halls, the forms & materials exhibition & leisure features of old and new buildings.

On the basis of "blank filling" and "integration", the design is intended to realize "new construction" through the overall integration of central axis space, architectural form and landscape creation.

The site of "Old Main Hall" is uplifted by 3m, which not only reverses the unfavorable situation that the original building was 3m lower than the urban roads, but also reduces the large volume earthwork filling and excavation for underground space. Besides, it creates favorable conditions for above-ground and underground circulation organization.

改扩建前 \ Before Reconstruction and Expansion

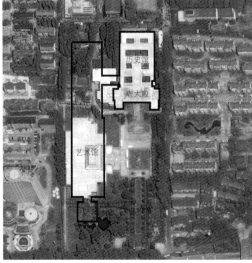

改扩建策略 \ Reconstruction Strategy

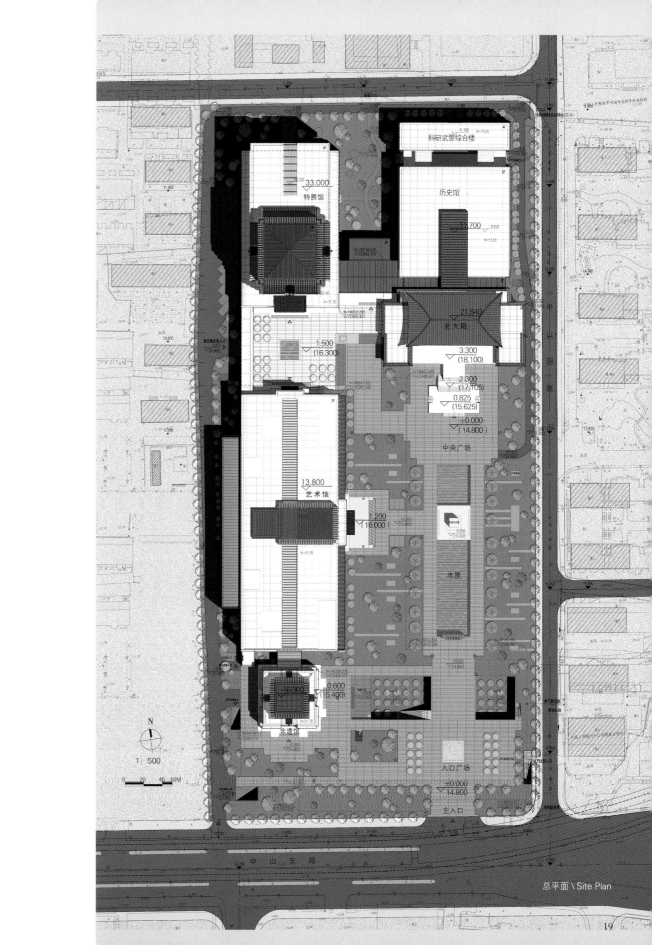

总平面 \ Site Plan

整个工程总建筑面积84900m²，包括保留的老大殿2000m²，局部改造并进行立面改造的艺术馆24000m²，其中地上建筑面积52910m²，地下建筑面积31940m²。二期工程要求对整个院域范围内所有建筑、设施、道路、环境进行整体规划设计。对历史馆仿辽式大殿按文物保护原则进行修缮，对原文物库房拆除，在此基础上还要整合文物库房。改扩建后的南京博物院新建了历史馆、特展馆、民国馆、非遗馆、数字化馆，并与改建的艺术馆形成了"一院六馆"的格局。

The gross floor area is 84,900m², including 52,910m² above ground and 31,940m² underground, also including 2000m² the reserved Old Mail Hall .The expansion project requires a overall planning and design of all buildings, facilities, roads, and environment within the range of the museum. It's necessary to renovate the imitation Liao-style Main Hall in the History Pavilion following the principles for historic preservation, demolish the original historical relics storehouse, and integrate the historical relics storehouse on the basis thereof. The renovated and expanded comprises the new History Pavilion, Special Exhibition Pavilion, the Republic of China Pavilion, Intangible Cultural Heritage Pavilion, Digital Pavilion and the reconstructed Art Gallery, and this layout is called "one museum with six pavilions".

老大殿、艺术馆及特展馆的空间组合 Space combination of Old Main Hall, Art Gallery and Special Exhibition Pavilion (Photographer: Chen Chang)

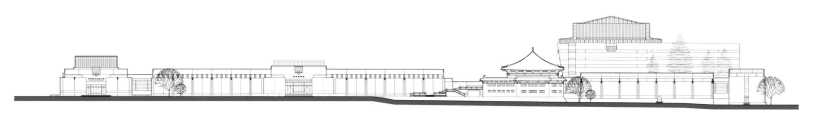

东立面 \ East Elevation

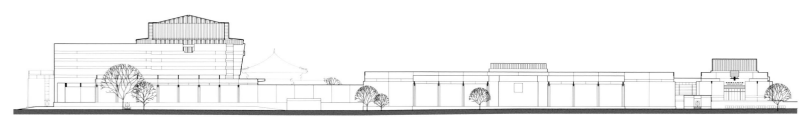

南立面 \ South Elevation

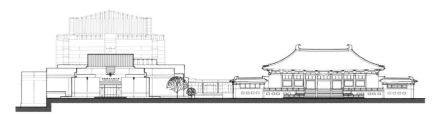

北立面 \ North Elevation

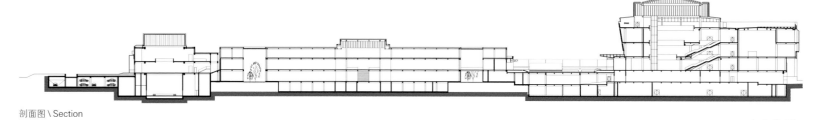

剖面图 \ Section

0 5 10 15M

中轴线景观 (张广源 摄影) \ Middle Axis Landscape (Photographer, Zhang Guangyuan)

由新建非遗馆看博物院中轴线空间(赵伟伟 摄影) \ Axle Space View from the Newly Built Intangible Cultural Heritage Gallery (Photographer: Zhao Weiwei)

　　南京博物院挑选的石材颜色为灰白基调，并且夹杂着类似火焰纹的暗红色的线和点，通过对石材表面做古典面处理，粗犷而内敛，温润而有厚重感。室内外还选用了紫铜板装修，铜板的质朴、庄重与典雅既与整体的设计氛围相吻合，又能与仿辽式老大殿的琉璃瓦屋顶取得协调。

The stone selected for Nanjing Museum whose colour is in a grey tone mingled with garnet lines and points that look like flame pattern; the classical surface treatment for stones contributes to the straight forward & implicit and mild & decorous nature. The interior and exterior are decorated with simple, unadorned, grave and elegant tough pitch copper boards that tally with the overall design atmosphere and harmonize with the glazed tile roof of imitation Liao-style old main hall.

艺术馆入口看老大殿（赵伟伟 摄影）\ View towards old Main Hall from the Entrance of Art Gallery (Photographer，Zhao Weiwei)

1 | 2

1 特展馆前广场看老大殿（赵伟伟 摄影）\ View towards Old Main Hall from the square in front of Special Exhibition Pavilion (Photographer, Zhao Weiwei)
2 艺术馆入口看老大殿（陈 畅 摄影）\ View towards Old Main Hall from the entrance of Art Gallery (Photographer, Chen Chang)

老大殿与历史馆结合部（张广源 摄影）\ Junction between Old Main Hall and the History Pavilion (Photographer: Zhang Guangyuan)

老大殿与历史馆接合部（张广源　摄影）\ Junction between the Old Main Hall and The History Pavilion (Photographer: Zhang Guangyuan)

扩建后的南博地下建筑面积达 3 万多平方米，设计将 4 个大小不一的下沉庭院和 12 个采光中庭及天窗通过精心穿插安排在了地上、地下建筑里，并且经过对采光口的位置、大小及以及遮阳形式的精心推敲，极大地解决了公共空间的自然采光与通风，并且使得空间极具感染力和戏剧性。

After the expansion, the underground floor area of Nanjing Museum is more than 30,000 square meters. The design incorporates 4 sunken courtyards in different sizes and 12 lighting atria and skylights carefully interspersed in above-ground and underground structures, while the careful determination of the location, size and shading form of daylight opening has brought about perfect natural lighting and ventilation for public space, and contributes to the highly catching and dramatic space.

非遗馆下沉庭院（陈　畅　摄影）\ Sunken Courtyard of Intangible Cultural Heritage Pavilion (Photographer，Chen Chang)

非遗馆全景（陈 畅 摄影）\ Intangible Cultural Heritage Pavilion (Photographer, Chen Chang)

艺术馆入口（张广源 摄影）\ Art Gallery Entrance (Photographer, Zhang Guangyuan)

艺术馆正立面（张广源 摄影）\ Front Elevation of the Art Gallery (Photographer, Zhang Guangyuan)

由艺术馆门廊看特展馆（陈 畅 摄影）\ View towards Special Exhibition Pavilion from Art Gallery (Photographer, Chen Chang)

特展馆东南角透视（陈 畅 摄影）\ Southeast Perspective of Special Exhibition Pavilion (Photographer, Chen Chang)

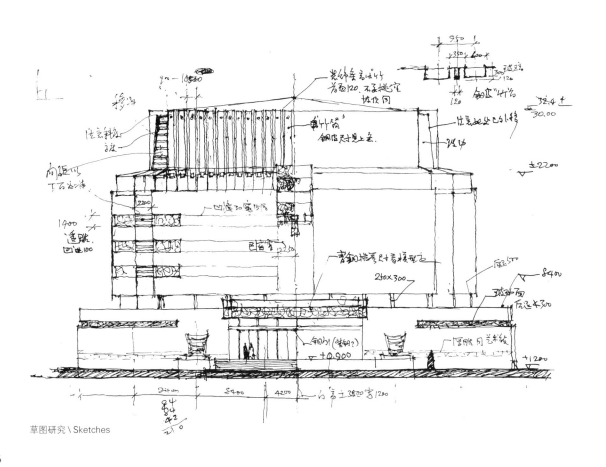

草图研究 \ Sketches

特展馆局部透视（赵伟伟 摄影）\ Special Exhibition Pavillion (Photographer : Zhao Weiwei)

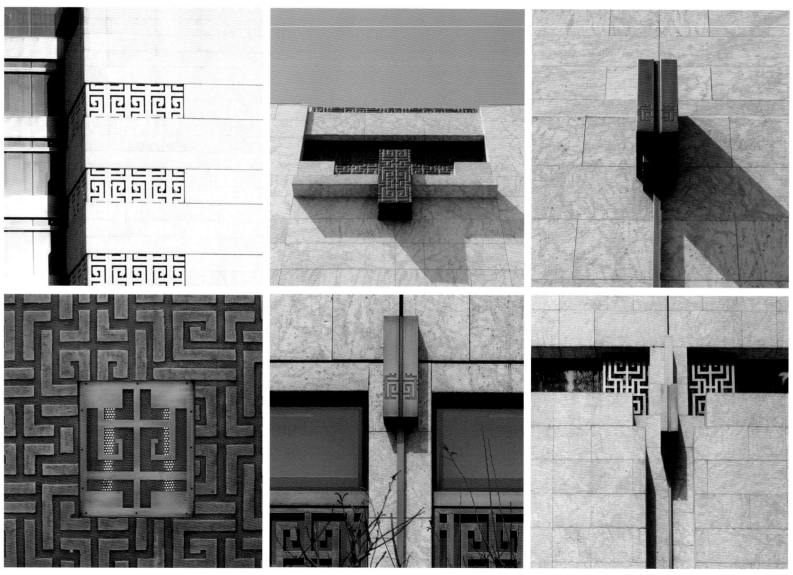

细部特写(赵伟伟 摄影)\ Details (Photographer, Zhao Weiwei)

艺术馆西立面（陈 畅 摄影）\ West Evelation of Art Gallery (Photographer: Chen Chang)

新设计的展馆室内中庭空间处理借鉴了老大殿"藻井"的手法，吊顶的梁格采用了紫铜板，形成了既简约现代又古朴典雅的现代建筑"藻井"，使得恢弘的大厅空间有了自己的个性，而且还调整了玻璃的进光量。

The newly designed indoor atrium space of exhibition hall borrows ideas from the "caisson" in old main hall.The grid beam made from tough pitch copper board in suspended ceiling constitutes the concise, stylish, unsophisticated and elegant modern "caisson" that offers a distinctive feature for the extensive lobby space; additionally, the light transmission rate of glass has been adjusted accordingly.

特展馆四楼（张广源 摄影）\ The Fourth Floor of Special Exhibition Pavilion (Photographer, Zhang Guangyuan)

特展馆中庭仰视（赵伟伟 摄影）\ Upward View of the Atrium of Special Exhibition Pavilion (Photographer, Zhao Weiwei)

特展馆中庭（陈 畅 摄影）\ The Atrium of Special Exhibition Pavilion（Photographer，Chen Chang）

民国馆长廊（赵伟伟 摄影）\ Long Corridor in the Republic of China Pavilion (Photographer, Zhao Weiwei)

历史馆中庭楼梯（赵伟伟 摄影）\ Staircase in the Atrium of History Pavilion (Photographer, Zhao Weiwei)

历史馆中庭 (赵伟伟 摄影) \ Atrium of History Pavilion (Photographer, Zhao Weiwei)

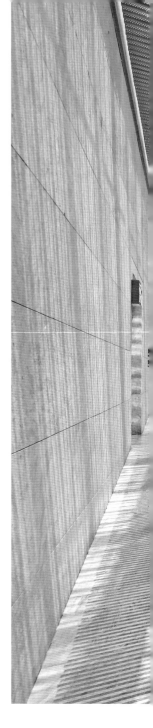

休息廊看老大殿（张广源 摄影）\ Viewing the Old Main Hall from the Veranda (Photographer, Zhang Guangyuan)

历史馆与老大殿之间的通道（张广源 摄影）\ Passageway between the History Museum and the Old Main Hall (Photographer, Zhang Guangyuan)

1 | 2

1 保留的原有建筑的木楼梯细部（张广源 摄影）
\ Wooden Staircase Detail of the Original Building Retained (Photographer, Zhang Guangyuan)

2 新老馆接合部的通廊中，保留了原有木楼梯。（张广源 摄影）
\ In the corridor at the junction between the new and old main halls, the original wooden stairs are retained. (Photographer, Zhang Guangyuan)

连接各馆的地下通道（张广源 摄影）\ Underground Passage Connecting the Museums (Photographer, Zhang Guangyuan)

连接各馆的地下长廊(张广源 摄影) \ Underground Corridor Connecting the Museums (Photographer: Zhang Guangyuan)

1 | 2 | 3
1 特展馆镇院之宝展厅（赵伟伟 摄影）\ Exhibition Hall of the Special Exhibition Pavilion for the Most Precious Treasures of The Museum (Photographer, Zhao Weiwei)
2-3 历史馆展厅（赵伟伟 摄影）\ Exhibition Hall of History Pavilion (Photographer, Zhao Weiwei)

龙泉青瓷博物馆
LONGQUAN CELADON MUSEUM

合作者 吴妮娜、杨 涛、刘鹏飞、李澍田、陈 悦
设 计 2007 / 竣 工 2012
Co-designers: Wu Nina, Yang Tao, Liu Pengfei, Li Shutian, Chen Yue
Design Time: 2007 / Completion Time: 2012

破土而出，生根自然
Break through the soil, and take roots in a natural manner

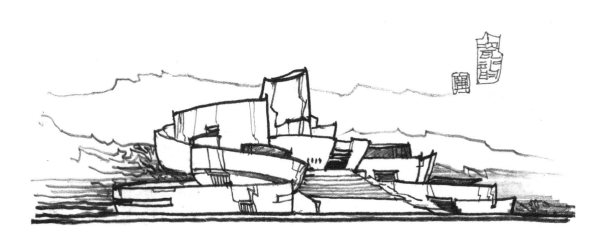

破土而出的"建筑"(赵伟伟 摄影)The "Building" Breaking Through the Soil (Photographer: Zhao Weiwei)

建筑从山体中"生长"出来(赵伟伟 摄影)\ The Building "Grows Out" of The Mountain (Photographer: Zhao Weiwei)

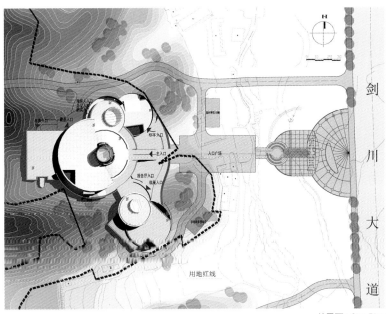

总平面 \ Site Plan

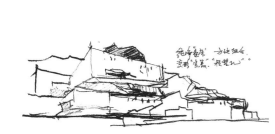

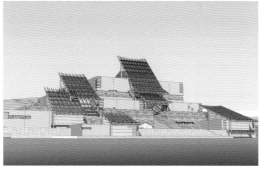
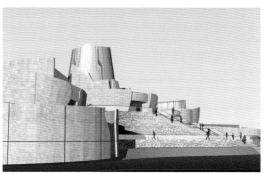

比较方案 \ Proposals

龙泉青瓷博物馆位于浙江省龙泉市主城区以南的城市入口处。占地面积6000m²，总建筑面积10000m²。基地地貌由两个平缓的山脊以及中间的洼地构成。背面山体给建筑提供了优美的环境背景。

由于历史上青瓷生产的高度发达，目前在龙泉四周的田野上随处可发掘到过去留下的窑址和青瓷碎片。处在这一特定的环境里，建筑以"瓷韵——在田野上流动"为创意，以一种非建筑的手法来表达这一博物馆的形象，如同考古发掘中层层叠叠的青瓷器物破土而出，自然地放置在田野之中。建筑造型是青瓷这一珍贵的文化遗产的抽象表达。

换一个角度去欣赏，它又像绘画艺术家笔下的一组景物，展现出了一幅恬静、优美的画面。

Longquan Celadon Museum, is located at the south entrance of the main urban area of Longquan, covering a land area of 6,000 m², and a gross floor area of 10,000 m². The site terrace is composed of two gentle ridges and a basin in between. The mountain at the back provides a picturesque environment for the property.

Due to the highly developed celadon production in history, kiln relics and celadon debris can be commonly excavated everywhere on the open fields in Longquan. Being put in this special context, the design has put "celadon rhythm flowing on the open field" as its development concept which presents the image of the museum in a non-architectural approach, just like celadon wares having been unearthed from an archeological excavation, naturally sitting on the open field. The architectural configuration of the museum is an abstract expression of this precious cultural heritage.

From another perspective, the museum also resembles an array of elements created by a painting artist, unfolding a tranquil and poetic image.

| 1 | 2 |
| 3 | 4 |

设计素材 \ Design Reference

1 田野中散落的匣钵和瓷器碎片 \ Saggars and Porcelain Fragments Scattered in Field
2 基地背景 \ Site   3 古龙窑 \ Dragon Kiln   4 青瓷 \ Celadon

草图研究 \ Sketches

1 庭院 \ Courtyard
2 办公 \ Office
3 教研室 \ Teaching & Research Room
4 机房 \ Machine Room
5 技术 \ Technology
6 编目 \ Cataloguing Room
7 鉴赏室 \ Appreciation Room
8 摄影 \ Photography
9 修复 \ Repairation Room
10 空调机房 \ Air Conditioning Machine Room
11 工具间 \ Tool Room
12 基本库房 \ Store Room
13 临时展厅 \ Temporary Exhibition Hall
14 管理间 \ Administration Room

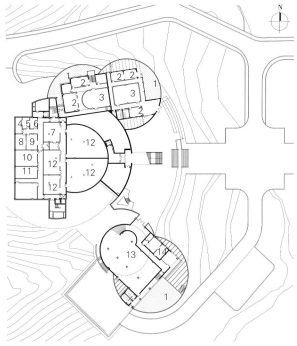

一层平面 \ Ground Floor Plan

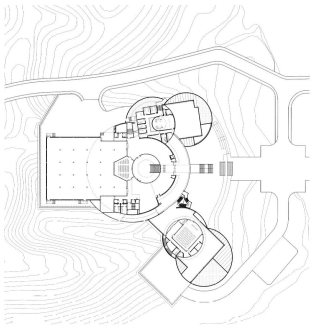

二层平面 \ Second Floor Plan

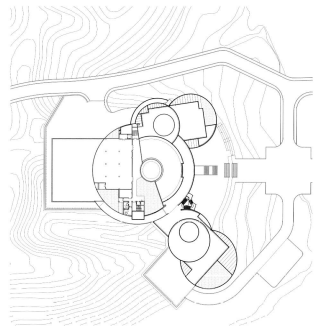

三层平面 \ Third Floor Plan

建筑师尝试以青瓷器物，匣钵为原型，经过抽象转换形成一种新的"语言"，即以双曲面的钵体单元和收分的圆形筒体相组合，来塑造建筑的整体形象。这些单元自由地镶嵌在这片坡地上，恰似沉睡在地下的青瓷器物破土而出，令人浮想联翩。

Designer tries to use the celadon as an original form, with the strategy of abstract transition, to form a new kind of "language", which takes both the battered round pails and the bowl-shape units to build the whole image of the building. These units are freely inlaid on the sloping site, as unearthed celadon relics, inspiring and delightful.

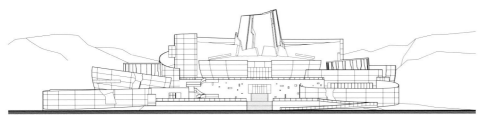

东立面 \ East Elevation

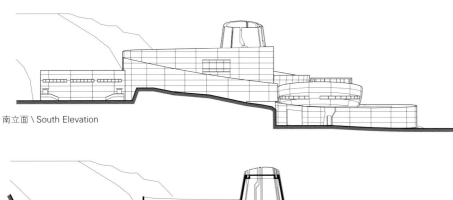

南立面 \ South Elevation

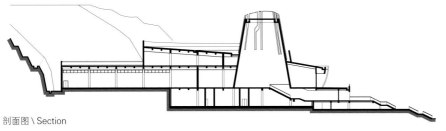

剖面图 \ Section

1　展厅 \ Exhibition Hall
2　放映室 \ Screening Room
3　序厅 \ Preface Hall
4　大厅 \ Lobby
5　设备 \ Equipment
6　基本库房 \ Storeroom
7　龙窑 \ Dragon Kiln

主入口全景（赵伟伟 摄影）\ Main Entrance (Photographer, Zhao Weiwei)

1	2
3	4

1 青瓷体块的楼梯间从混凝土外墙探出来（陈 畅 摄影）\ Celadon Block Shaped Staircase is Extending Out of the Concrete External Wall (Photographer, ChenChang)

2 嵌入清水墙的"瓷片"（赵伟伟 摄影）\ "Crackle china" Embedded in Bare Concrete Wall (Photographer, Zhao Weiwei)

3 弧形清水墙围合的办公内院（赵伟伟 摄影）\ Office Garth Surrounded by Arc-shaped Bare Concvete Wall (Photographer, Zhao Weiwei)

4 建筑与远山（赵伟伟 摄影）\ Building and Distant Mountain (Photographer, Zhao Weiwei)

建筑融入环境之中（赵伟伟，摄影）/ Building Integrated into Environment (Photographer: Zhao Weiwei)

1 | 2

1 至二层观景平台的坡道（赵伟伟　摄影）\ The Ramp Leading to the Viewing Platform on the Second Floor (Photographer, Zhao Weiwei)
2 清水混凝土与文化石的对比（赵伟伟　摄影）\ Comparison Between Bare Concrete and Art Stone (Photographer, Zhao Weiwei)

墙体采用清水混凝土，大片暖灰色调与出窑后的匣钵相似，与点缀其间的青绿色的瓷筒片断，以及象征窑体的略显粗犷的文化石基座相互组合，色彩及材料质感浑朴自然。

立面上的瓷坯碎片、变形的门洞、散乱的投柴孔，似乎留下了些许历史的印迹，它隐喻青瓷的新生，也再现了我们希望表达的建筑与自然共生的田园意境。

Wall are made of bare concrete. Large surface of warm grey tone which looks like fired saggar is combined with dotted blue green porcelain tube fragment and cultured stone pedestal which looks a little rough. Color and material texture is simple and natural.
Porcelain base fragment on façade, deformed door way and scattered firewood seem to have left some historical print. It symbolizes rebirth of celadon and represents our expectation for rural prospect where buildings coexist with nature.

3 | 4

3 楼梯一角（赵伟伟 摄影）\ Part of the Staircase (Photographer: Zhao Weiwei)
4 文化墙上的"投柴口"窗洞（赵伟伟 摄影）\ "Firewood Holes"–Shaped Window Opening in The Art Wall (Photographer: Zhao Weiwei)

| 1 | 2 | 3 1-3 庭院内景（1、3 陈 畅 摄影，2 赵伟伟 摄影）\ Inner view of Garden (Photographer, 1、3 Chen Chang, 2 Zhao Weiwei)

博物馆的入口设在标高 –10.0m 处，观众通过长长而低矮的"龙窑"甬道进入标高为 +0.00m、高度达 21m 的圆筒形序厅。甬道墙面仿窑壁杂色流釉，序厅墙及顶面均为青灰色，空间形态及色彩的对比给人以强烈印象。观众仿佛由"龙窑"窑床进入青瓷器物之中。序厅的光线由顶部"裂缝"中洒入，形成独特的抽象图案效果。

The Museum entrance is located at an elevation of -10.0metres. Visitors travel through a long and low "dragon kiln" path before entering a 21 metres high cylindrical hall at an elevation of +0.00metre. The surface material along path imitates mixed color sagging. Lobby wall and top are steel grey. The Spatial form and color contrast impress people intensively. Visitors feel as if they walked into blue glazed porcelain through "dragon kiln" floor. Light ray sprinkles into lobby through "crevice" on the top to form unique abstract pattern.

光线从序厅顶部的"裂缝"中洒入（赵伟伟 摄影）
\ The Light Passes Through the "Fissure" On the Top of Preface Hall (Photographer, Zhao Weiwei)

龙窑甬道（赵伟伟 摄影）\ Dragon Kiln Corridor (Photographer: Zhao Weiwei)

龙窑内仿窑壁杂色流釉（赵伟伟 摄影）\ Imitation Kiln Wall Variegated Sagging in Dragon Kiln (Photographer: Zhao Weiwei)

中国海盐博物馆
CHINA MUSEUM OF SEA SALTS

合作者 程跃文、吴妮娜、杨 涛、李澍田、吴文竹
设 计 2007 / 竣 工 2009
Co-designers: Cheng Yuewen, Wu Nina, Yang Tao, Li Shutian, Wu Wenzhu
Design Time: 2007 / Completion Time: 2009

品体之美和滩涂广阔的意境结合起来
Combination of the image of the sea salt crystals and the wide stretch of beach

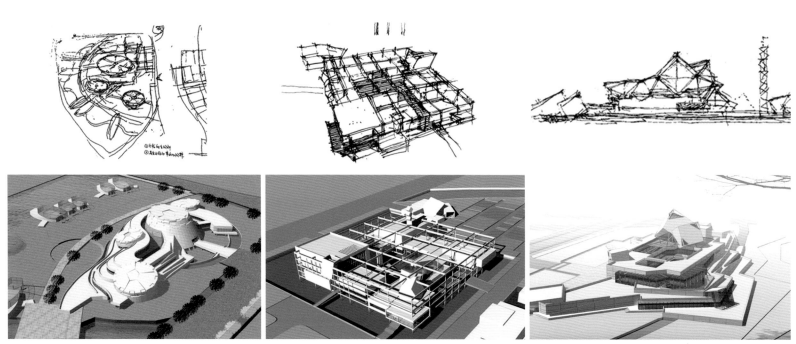

比较方案 \ Proposals

中国海盐博物馆位于江苏省盐城市，基地位于贯穿盐城市的重要河流串场河以东，总建筑面积17800m²。

建筑造型是海盐结晶体的演绎，广阔的海边滩涂为海盐的生产提供了独特的环境，如何把这些元素融入到建筑设计之中，是我们所探索的课题。

旋转的晶体与层层跌落的台基相组合，就像一个个晶体自由地洒落在串场河沿岸的滩涂上，造型独特。

China Museum of Sea Salts (Yancheng, Jiangshu), with a total construction area of 17,800 m², has its foundation to the east of Chuanchang River, a key river flowing through Yancheng.

The configuration of the museum is a presentation of sea salt crystals. The extensive sea beach provides the unique environment for the production of sea salts. And how to incorporate these elements into the design is at the forefront of our research agenda.

The concept behind the rotary crystals perching on a platform in a gradient is to recreate the unique scene of various salt crystals freely distributing on the beach by Chuanchang River.

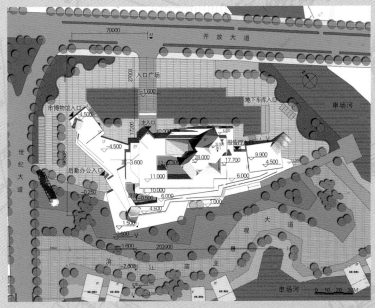

总平面 \ Site plan

结合台基造型和平面功能，创造一个功能合理、内外空间灵活别致的建筑。开放的广场，以及临串场河观景带形成开放的建筑场所，观众在不知不觉之中，被引导至各个观景平台，几千年煮海文明与现代文明交融汇聚的场景尽收眼底。

Create a building with rational function deployment and flexible & unique inner and outer spaces based on foundation feature and 2-dimensional function. Establish an open architectural place by combining the open square and the viewing zone approaching the Chuanchang River. Visitors are unconsciously led to viewing platforms, where they have a panoramic view of the fusion of thousands of years of the civilization with modern civilization.

东南立面 \ Southeast Elevation

西北立面 \ Northwest Elevation

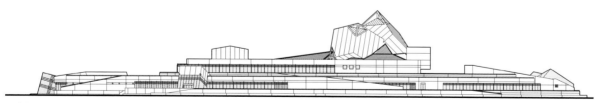

西南立面 \ Southwest Elevation

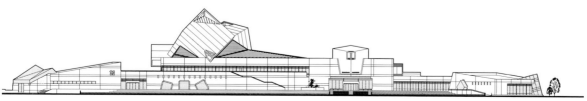

东北立面 \ Northeast Elevation

主立面全景（赵伟伟 摄影）\ View of Main Elevation (Photographer, Zhao Weiwei)

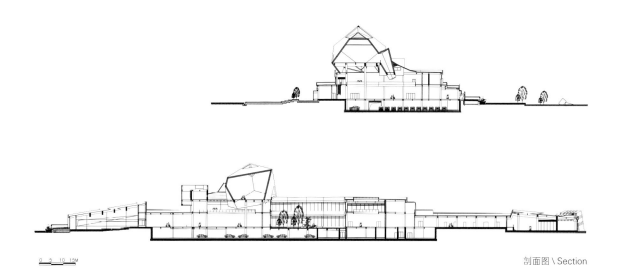

剖面图 \ Section

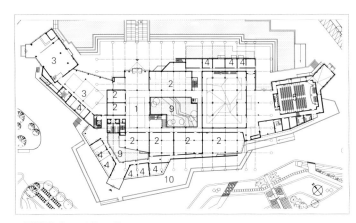

一层平面 \ Ground Floor Plan

1	中庭 \ Atrium
2	展厅 \ Exhibition Hall
3	展厅（独立陈列）\ Exhibition Hall (Individual)
4	办公 \ Office
5	报告厅 \ Lecture Hall
6	纪念品 \ Souvenir Shop
7	阅览室 \ Reading Room
8	咖啡吧 \ Cafe Bar
9	内院 \ Courtyard
10	室外平台 \ Outdoor Platform

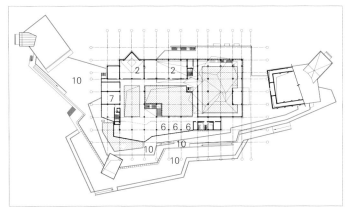

二层平面 \ Second Floor Plan

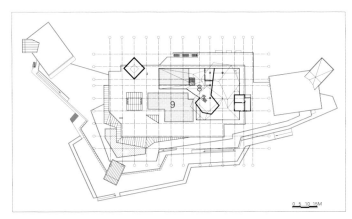

三层平面 \ Third Floor Plan

入口透视（赵伟伟 摄影）\ Entrance (Photographer, Zhao Weiwei)

串场河方向全景（赵伟伟 摄影）
\ Panorama from the River Direction (Photographer，Zhao Weiwei)

外景（赵伟伟 摄影）\ Exterior View (Photographer, Zhao Weiwei)

外景（赵伟伟 摄影）\ Exterior View (Photographer, Zhao Weiwei)

细节及材料选用十分重要。暖灰色砂岩外墙，质感肌理贴近自然，顶部晶体采用银白色金属铝板镶嵌抛光不锈钢钉，通过多角度的反射，在阳光的照射下熠熠生辉，并减弱了铝板面可能积灰的印象。值得一提的是，石材的颜色、质感以及金属钉的大小都曾在现场经过多次挂样比较，才得以实现这一效果。

　　建筑标准控制适当，建筑工程每平方米造价3400元，为同类工程最低。

Detaied design and material selection are very important. External wall of warm grey sandstone has nature approaching texture. Crystal on its top is decorated by silver color aluminum plate embedded with polished stainless steel nail, which sparkles in sunshine through multi-angle reflection and diminish the impression that aluminum plate accumulates dust. It is worth to mention that color and texture of stone material as well as size of metal nail have experienced sample comparison at the site for many times in order to achieve the current effect.

With properly controlled architectural standard, construction and installation cost per meter is 3400RMB, which is the lowest among similar projects.

局部（赵伟伟　摄影）\ Local (Photographer, Zhao Weiwei)

西南角入口（赵伟伟 摄影）\ Entrance at Southwest Corner (Photographer, Zhao Weiwei)

博物馆内部围绕中庭展开，共分两层，设立"生命之旅"、"史海盐踪"、"煮海之歌"、"盐与盐城"四个展厅。通过形体的穿插，自然形成两大庭院，分别供游客和办公人员观赏休憩。布展采用二、三维结合的形式，增加了更多的体验和互动，中庭停泊的木船、场景大厅的实景还原，观众还可通过架在实景中的桥近距离感受盐的产销过程。

Deployed around atrium, the interior of museum is composed of two floors and consists of four exhibition halls, i.e. "Journey of Life", "Salt in the History", "Song of Sea-cooking", and "Salt and City". The interpenetration of bodies naturally brings about two courtyards where visitors and office staff could enjoy the sight or have a rest. Through the combination of 2D and 3D, the exhibition layout presents more visitor experience and interactions; thanks to the wooden boat in atrium and the virtual reproduction in the scene hall, the visitors could learn about the salt production and marketing processes on site at the bridge erected in the scenery.

内景（赵伟伟 摄影）\ Interior view (Photographer, Zhao Weiwei)

场景演示厅内景（赵伟伟　摄影）\ Interior of Scenario Demonstration Hall (Photographer, Zhao Weiwei)

宁夏大剧院
NINGXIA THEATER

花开盛世
A Blossom in Flourising Age

合作者 郑庆丰、唐 晖、程跃文、陈 悦、段继宗、叶 俊、杨 涛、骆晓怡、刘鹏飞、潘知钰
设 计 2008 / 2014 基本竣工

Co-designers: Zheng Qingfeng,Tang Hui,Cheng Yuewen,Chen Yue,Duan Jizong,Ye Jun,Yang Tao,Luo Xiaoyi,Liu Pengfei,Pan Zhiyu
Design Time: 2008 / Basically Completion Time: 2014

全景鸟瞰（右上：图书馆 左下：博物馆）(陈 畅 摄影) \ Panoramic Aerial View (Up-right:Library, Bottom-left:Museum) (Photographer, Chen Chang)

草图研究 \ Sketches

地处银川城市核心区人民广场东侧，南临北京路，与文化艺术中心、宁夏博物馆、图书馆围合成东组团广场。剧院建筑总面积 4.9 万 m^2。

由于博物馆及图书馆均为方整的造型，大剧院采用了外方内园的图式，并通过工作模型推敲了三者的体量关系，使之成为一个完整的组合，并将作为主体的大剧院的形象突显出来。

The project is located to the east of People's Square in Yinchuan Core District, with Beijing Road to the south, and the Culture and Art Center to the north. With the newly-built Ningxia Museum and Ningxia Library to its west, they together enclose the east cluster square. The theater spreads over a total area of 49,000 m^2.

As the museum and the library all take a square shape, the theater features an outer square and inner circle design. Much study and exploration had been made on the volume of the three structures through mockup tests to find an integrated combination and to emphasize the image of the theater as a main part.

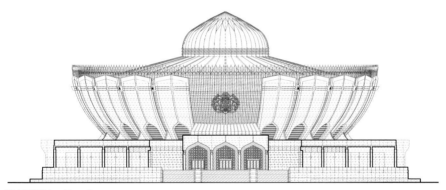

西立面 \ West Elevation

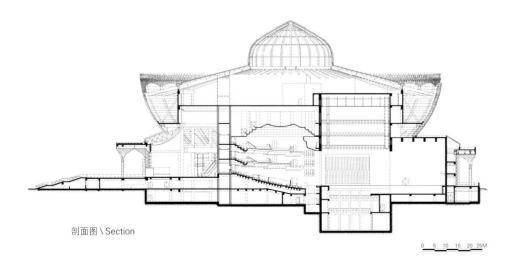

剖面图 \ Section

1 主舞台 \ Main Stage
2 侧舞台 \ Stage
3 后舞台 \ Back Stage
4 多功能厅 \ Multi-Functional Room
5 休息厅 \ Rest Room
6 多功能厅前厅 \ Anteroom of Multi-Functional Room
7 送风静压箱 \ Plenum Chamber
8 汽车库 \ Garage
9 绘景间 \ Painting Room
10 化妆间 \ Dressing Room
11 前厅 \ Anteroom
12 内庭院 \ Inner Contyard
13 观众厅池座 \ Loge of Auditorium
14 休息厅 \ Rest Room
15 办公 \ Administration Room
16 机房 \ Equipment Room
17 主舞台台仓 \ Understage of Main Stage

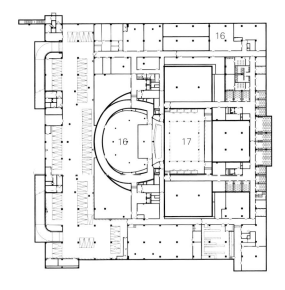

地下层平面 \ Basement Floor Plan

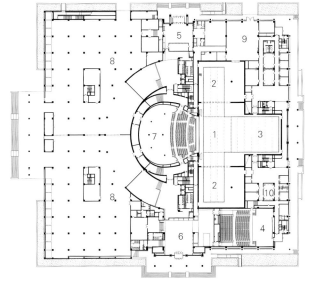

一层平面 \ Ground Floor Plan

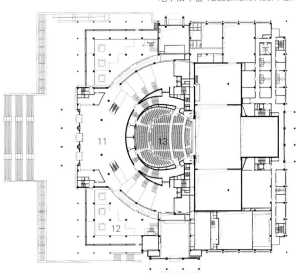

二层平面 \ Second Floor Plan

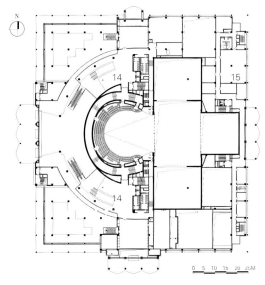

二层平面 \ Third Floor Plan

银川是宁夏回族自治区首府，宁夏的地域特色伊斯兰文化的传承是我们所重视的，因为这使建筑的"唯一性"成为可能。我们希望宁夏大剧院能够与北京、上海、广州或在其他一些地方的大剧院区分开来。当然，不言而喻，地域文化的表达不是简单的移植而是在全球化背景下的再创作。经过抽象、提炼、最后升华为符合工业化生产方式和现代人审美理想的建筑意象。

"花开盛世"，较恰当的表达了我们对"现代的，中国宁夏的"的设计创意。

Yinchuan is the capital of Ningxia Hui Autonomous Region, and we have put great value on the inheritance of the regional feature with Ningxia Islamic culture. Because that makes the architecture's uniqueness possible. We hope that Ningxia Theater has its own special elements that are distinctive from other theaters in Beijing, Shanghai, Guangzhou, and other places. It goes without saying that the expression of regional culture is not just a simple transplanting but the recreation in the context of globalization. An innovative design work is produced through abstraction, extraction, and the ultimate sublimation of the structural image that is in conformance with the industrialized mode of production and the modern aesthetic ideal.

"A blossom in flourising age" gives an appropriate presentation of our design concept of "Modern and Ningxia of China".

西立面全景透视(陈 畅 摄影) \ Perspective Panoramic View of West Facade (Photographer: Chen Chang)

西南方向外景（陈 畅 摄影）\ Southwest Perspective (Photographer, Chen Chang)

西南方向外景（陈 畅 摄影）\ Southwest Perspective (Photographer: Chen Chang)

东面临街外景（陈 畅 摄影）\ East Frontage Perspective (Photographer, ChenChang)

南面临街外景 (陈 畅 摄影) \ South Frontage Perspective (Photographer, ChenChang)

屋面细部（陈 畅 摄影）\ Roofing Detail (Photographer, ChenChang)

鸟瞰(陈 杨 摄影) \ Aeiral View (Photographer, ChenChang)

入口门廊细部（陈 畅 摄影）\ Entrance Porch Detail (Photographer, ChenChang)

西（主）入口外景 (陈 畅 摄影) \ West (Main) Entrance Exterior (Photographer: Chen Chang)

门廊细部 (陈 畅 摄影) \ Porch Detail (Photographer, ChenChang)

入口环廊细部(陈 畅 摄影) \ Entrance Surrounding Corridor Detail (Photographer, ChenChang)

细部特写(陈 畅 摄影)＼Details (Photographer, ChenChang)

细部特写(陈畅 摄影)\Details (Photographer: ChenChang)

观众厅内景（陈 畅 摄影）\ Auditorium Interior (Photographer, ChenChang)

原设计意图是外立面及大厅以纯净的白色为主调，观众厅则打破通常"三块板"模式，采用装饰感较强、地域特色浓烈的穹顶天花，形成对比。

浙江大学张三明教授（声学设计）以及舞台灯光设计单位给予了技术支持。

Orginally, white was designed to be the dominant color for the facades and hall, while the auditorium has broken the usual "three plates" mode, adopting a strong decorative dome ceiling, with strong regional characteristics, to form the contrast.

Professor Zhang Sanming (acoustic design) of Zhejiang University and stage lighting design institute have provided technical support for the project.

原设计效果图 \ Original Design Drawing

室内大厅实景 (陈 畅 摄影) \ View of Interior Lobby (Photographer，ChenChang)

1.中央大厅内墙原为白色石材（已选定样板），并以金线装饰，既可体现地域文化，又可使空间显得比较纯净。但最后代建单位改为中灰带花纹的石材，室内效果杂乱。

2.大厅主墙面（左图）原设计考虑为浅浮雕，由下而上逐步隐退，以保持大厅空间的整体性。但最后主管单位决定用色彩强烈的壁画（右图），不大的室内空间更显局促。

3.以上问题虽经过多次提出意见，但最后仍未能解决，遗憾！

1. Originally, the interior wall of the central hall was white stone (sample selected), and decorated with gold thread, which could both reflects the local culture, and also make the space more pure. However, the agent construction finally changed it into gray stone with pattern, resulting in messy interior effect.

2. The original design of the main wall in the hall (left picture) considered for bas-relief, with gradual retreat from the bottom up, so as to maintain the integrity of space in the hall. However, the unit in charge finally decided to apply mural with strong color (right picture), making the interior space which is not large even more cramped.

3. Although the above problems had been proposed for many times,problems were still not solved in the end,regret!

中国港口博物馆
CHINA PORTS MUSEUM

合作者 陈 玲、刘翔华、叶 俊、张朋君
设 计 2011 / 竣 工 2014
Co-designers: Chen Ling,Liu Xianghua,Ye Jun,Zhang Pengjun
Design Time: 2011 / Completion Time: 2014

隐喻，在抽象与具象之间的挣扎
Metaphor is the struggle between the abstract and concrete

夜景（陈畅 摄影）\ Night view (Photographer: Chen Chang)

西南方向透视 (陈 畅 摄影) \ Southwest Perspective (Photographer, Chen Chang)

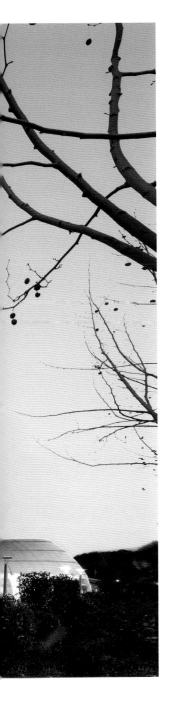

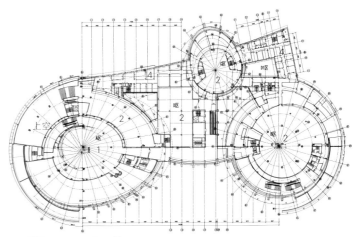

二层平面 \ Second Floor Plan

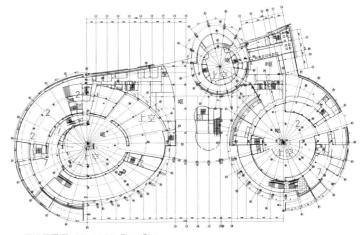

一层夹层平面 \ Mezzanine Floor Plan

1 中庭 \ Atrium
2 展厅 \ Exhibition Hall
3 大厅 \ Lobby
4 办公 \ Office
5 报告厅 \ Lecture Hall
6 库房 \ Reading Room
7 休息厅 \ Cafe Bar

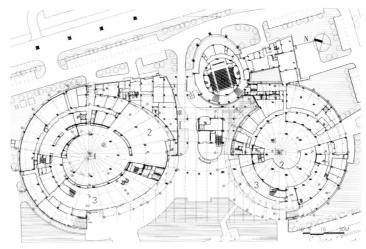

一层平面 \ Ground Floor Plan

设计立意突出海洋文化，以颇富现代气息的非线性造型和体现现代科技的外墙材料表现宁波北仑港锐意创新的发展形象，力图打造一个有特色的文化休闲场所，强化它对公众的吸引力，以带动新区的开发。从建成后的情况看，已基本达到了这一要求。

The design concept highlights ocean culture and presents the Ningbo Beilun Port's innovative image through extremely stylish nonlinear structure and the metal sheets that incarnate modern technologies, trying hard to build it into a culture and leisure space with strong public character so as to intensify its appeal to the public and spur on the development of the new town. Seen from the current situation after completion, the building has basicly archived the goal.

西北方向透视（陈 畅 摄影）\ Northwest Perspective (Photographer, Chen Chang)

局部（陈 畅 摄影）\ Local (Photographer, Chen Chang)

室内空间（陈 畅 摄影）\ Interior Space (Photographer, Chen Chang)

室内设计注重简洁，流畅。风格与外观统一，同时控制铝板、石材的使用面积，力求降低造价。

The interior design mainly aims at simplicity and fluency. Style and appearance are unified and area of aluminum plate and stone is controlled to reduce cost.

室内空间（陈 畅 摄影）\ Interior Space (Photographer, Chen Chang)

室内空间（陈 畅 摄影）\ Interior Space (Photographer: Chen Chang)

室内空间（陈畅 摄影）\ Interior Space (Photographer, Chen Chang)

湘潭城市规划展览馆及博物馆
XIANGTAN URBAN PLANNING EXHIBITION HALL AND MUSEUM

合作者　王大鹏、柴　敬、王禾苗、杨思思、胡晓明、叶　俊
合作单位　湘潭市建筑设计院
设　计　2009＼2014 基本竣工
Co-Jesigners: Wang Dapeng,Chai Jing,Wang hemiao,Yang sisi,Hu Xiaoming,Ye Jun
Cooperation Company: Xiangtan City Architectural Design Institute
Des gn Time: 2009＼Basically Completion Time: 2014

黑白 · 光影
Black Vs· White,Light Vs· Shadow

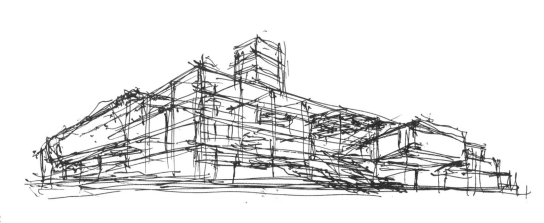

西南方向全景透视（陈 畅 摄影）\ Southwest Panoramic Perspective (Photographer: Chen Chang)

项目总建筑面积38946㎡，其中地下4804㎡，地上建筑面积34142㎡。建筑物位于城市新区核心地带，紧临行政中心和新区梦泽湖景观，交通便捷，环境优美。主要由博物馆、规划展示馆、规划局办公楼三部分组成。设计创意立足于地理环境与人文历史，"韶山红"的基座、灰白色的墙面，特别是黑色的构架通廊分隔、同时又使三幢建筑连串整体，也形成了与环境相结合的公共空间。

The project has a gross area of 38946m², including 34142m² above ground and 4804m² underground. The building is located in the core zone of the new urban area, and close to the administrative center and the Mengze Lake landscape of the new area, featuring convenient transportation and graceful environment. It consists of mainly three parts: the Museum, the Planning Exhibition Museum, and the planning bureau office building. The design idea sets foothold on the geographical environment and the human history, the Shaoshan Red pedestal, the off-white walls, especially the black architecture corridor separation. Meanwhile, the three buildings are connected into one overall architecture, and also form into public space in combination with the environment.

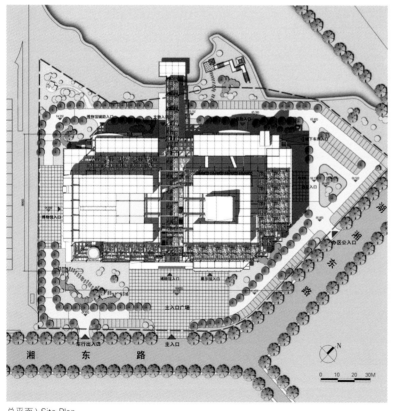

总平面 \ Site Plan

梦泽湖方向透视（陈 畅 摄影）\ View from Mengze Lake (Photographer, Chen Chang)

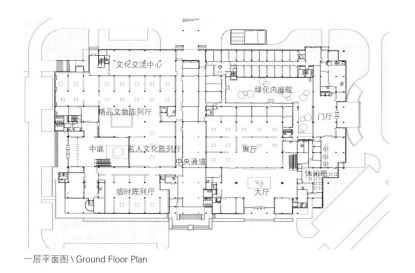

一层平面图 \ Ground Floor Plan

二层平面图 \ Second Floor Plan

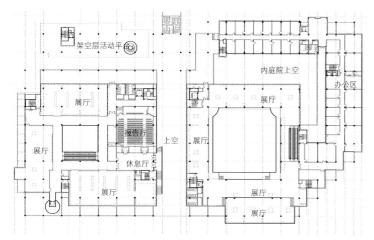

三层平面图 \ Third Floor Plan

四层平面图 \ Fourth Floor Plan

西南方向（左为博物馆入口）透视 (陈 畅 摄影) \ Southwest Perspective (Right to Museum Entrance) (Photographer, Chen Chang)

东南方向（规划展览馆入口）透视（陈 畅 摄影）
\ Southeast Perspective (Entrance to the Planning Exhibition Hall) (Photographer, Chen Chang)

规划馆主入口透视（陈 畅 摄影）\ Perspective of Main Entrance to Planning Exhibition Hall (Photographer，Chen Chang)

东立面（规划局办公楼）透视 (陈 畅 摄影) \ East Elevation Perspective (Office Building of the Planning Bureau) (Photographer, Chen Chang)

构架通廊入口（陈 畅 摄影）\ Architecture Corridor Entrance (Photographer, Chen Chang)

平面为"T"形的构架通廊既可连接南面的城市道路与北面的梦泽湖公园,也可连接东西两馆,供市民俯瞰梦泽湖公园(陈 畅 摄影)
\ The "T" shaped structure corridor, is connecting both the urban roads on the south and the Mengzehu Park on the north, also connects the east and west museums while rendering place for the public to overlook Mengzehu Park. (Photographer: Chen Chang)

设计通过对空间品质的塑造和细节的雕琢，力图营造一个展品与环境、人与环境、人与展品以及人与人之间的互动交流的和谐场所。

在白石老人的故乡湘潭齐白石纪念馆内，我们看到了一代宗师的巨幅篆刻。震撼之余，它也成为建筑中的一个构成元素，强化了墙面上的光影变化。构架在墙面上形成光影，建筑显得灵动而富于变化。

Through the design for textural space shaping and details, designers are trying to create a positive communicating venue for exhibit and environment, people and environment, people and exhibit and also between men.

In the Qibaishi Memorial Hall (situated at Xiangtan, the hometown of Qi Baishi), we are shocked at seeing the huge seal-cutting masterpiece of the great painter. The masterpiece itself also becomes a constituent element of the building, and strengthens the change in light and shadows on the wall.

1 构架及纹饰光影（陈 畅 摄影）\ Shadow of Architectural Structure and Emblazonment (Photographer, Chen Chang)
2 北部通廊东视（陈 畅 摄影）\ East Perspective of North Corridor (Photographer, Chen Chang)
3 局部透视（程泰宁 摄影）\ Detailed Perspective (Photographer, Cheng Taining)
4 西北角立面透视（陈 畅 摄影）\ Northwest Elevation Perspective (Photographer, Chen Chang)

湖州南浔行政中心
NANXUN ADMINISTRATIVE CENTER, HUZHOU

合作者 薄宏涛、于 晨、陈 玲、刘翔华、张朋君
设 计 2007 / 竣 工 2011
Co-cesigners: Bo Hongtao, Yu Chen, Chen Ling,Liu Xianghua,Zhang Pengjun
Design Time: 2007 / Completion Time: 2011

换一种思路，创造一个公众参与的开放空间
A different viewpoint, Create open space for public participation

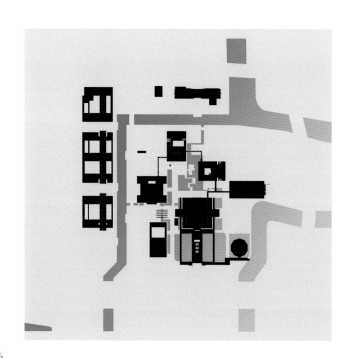

外景透视（曹 杨 摄影）/ Outdoor Perspective (Photographer: Cao Yang)

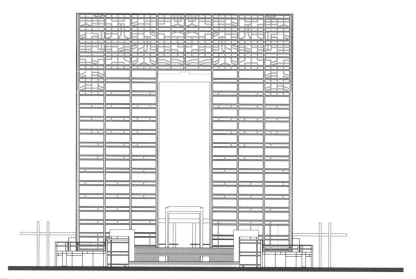

南立面 \ South Elevation

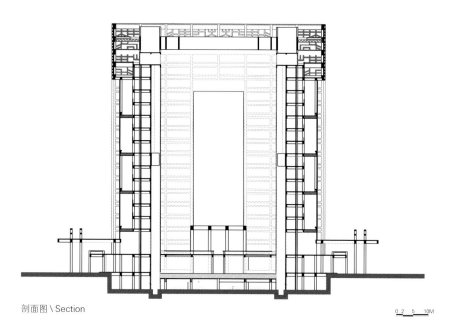

剖面图 \ Section

0 2 5 10M

主入口（赵伟伟 摄影）\ Main Entrance (Photographer, Zhao Weiwei)

水院景观(曹 扬 摄影)\ View of Water Courtyard (Photographer, Cao Yang)

底层庭院透视(赵伟伟 摄影)\ Ground Courtyard Perspective (Photographer: Zhao Weiwei)

二层景观平台（赵伟伟 摄影）\ Platform on the Second Floor (Photographer, Zhao Weiwei)

南浔古镇历史久远，文化底蕴深厚。如何使南浔江南水乡文化与行政建筑开放庄重的气质完美结合，是设计中的核心问题。

当下之所谓"新中式建筑"，基本仍然停留在手法主义的层面上。实际上，传统建筑空间更加注重一种与建筑功能、审美相联系的意境表达，而且，中国传统建筑基本上不存在单体建筑的概念，侧重的是群体建筑空间体系的建立。同时一个空间体系又包含在一个更大的体系之中，由单体到群体其实是一个"自相似性"的结构，建筑单体与群体的设计概念和寓意其实是同根同源的。

Nanxun Ancient Town enjoys a long history and profound culture. The core issue of the design is to find a solution to organically integrate Nanxun Jiangnan River Town culture and the open yet solemn quality the Administration Center presents. The so-called "Neo-Chinese Architecture" is in essence still staggering at the layer of expressionism. In fact, traditional architectural space puts more emphasis on the expression of an artistic conception that seeks an association and combination of architectural function and aesthetic function. Moreover, in the vocabulary of traditional Chinese architecture there is no such concept as "single building", and considerably big emphasis has been devoted on the concept of a cluster building spatial system. Even a spatial system is encompassed in a larger system. A cluster building is in fact a system which includes a bunch of self-similar structures. The design concept and meaning of single building and cluster building is essentially the same.

二层礼仪入口（赵伟伟 摄影）
\ Ceremonial Hall Entrance on the Second Floor (Photographer, Zhao Weiwei)

院落萦回，空间环绕，玻璃、钢构等一系列现代材料的引入又使建筑平添了一种生机盎然的时代感。中国园林的意境就在"似与不似之间"如画卷般展开……

The courtyards and spaces, as well as the introduction of a series of modern elements such as glass and steel components have added a sense of vitality and modernity to the building. The artistic conception of Chinese gardens is implicitly presented like unfolding a picture scroll....

1 2 | 3 4

1-4 庭院空间（1-3 赵伟伟 摄影，4 曹扬 摄影）
\ Courtyard (Photographer, 1-3 Zhao Weiwei, 4 Cao Yang)

湖南昭山两型发展中心

ZHAOSHAN TWO TYPE INDUSTRIAL DEVELOPING CENTER, HUNAN

合作者 王大鹏、沈一凡、孟 浩、汤 焱、祝 容、裘 昉
合作单位 湘潭市建筑设计院
设 计 2012 / 2014 基本竣工

Co-designers: Wang Dapeng, Shen Yifan, Meng Hao, Tang Yan, Zhu Rong, Qiu Fang
Cooperation Company: Xiangtan City Architectural Design Institute
Design Time: 2012 / Basically Completion Time: 2014

与自然共生，与文脉相承，与时代同步

Co-exist with nature, inherit traditional culture and keep pace with era

全景鸟瞰(陈畅 摄影) \ Panoramic Aerial View (Photographer: Chen Chang)

1 门厅 \ Foyer
2 中庭 \ Atrium
3 内院 \ Courtyard
4 办公 \ Office
5 会议厅 \ Conference Hall
6 展厅 \ Exhibition Hall
7 餐厅 \ Restaurant
8 银行营业厅 \ Banking Hall
9 政务大厅 \ Government Affairs Hall

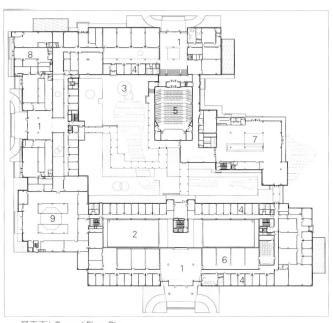

一层平面 \ Ground Floor Plan

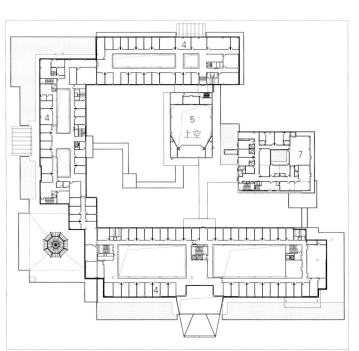

二层平面 \ Second Floor Plan

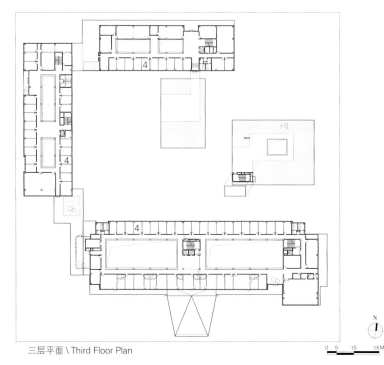

三层平面 \ Third Floor Plan

项目总建筑面积51781 m²，其中地下15791 m²，地上建筑面积35984 m²。距潇湘八景的昭山不远，建筑环山而建，建筑形体逐层跌落，犹为山体的延续，同时又形成了与自然环境相融无间的内部庭院。空间布局既体现了办公建筑的公共性，又营造了寄情山水的园林意境，单体造型表现传统建筑的水平构线和湖南"穿斗式建筑的细节"，经过提炼变形，试图塑造一个典雅精致，又颇有力度感的建筑形象。中国而现代，是我们的创作定位。本项目也为昭山两型示范区的建设奠定了基调。

The project has a gross area of 51781m², including 15791m² above ground and 35984m² underground. This project is near Shaoshan Mountain, which is one of eight famous sceneries in Hunan Province. The building is surrounded by mountains, so we make the form falls by story and looks like extension of the mountain volumn. Meanwhile, it forms internal courtyard well integrated into natural environment. The space inside not only represents public character of office building, but also creates garden prospect enchanted in landscape. Single modeling represents horizontal outline of traditional architecture and "column-and-tie architecture detail". Its extraction and transformation attempt to create an elegant, delicate and strong architectural image. Our goal is to mix the Chinese characteristic with the modern form. This project also established a example for construction of Two Type Industrial Sample area in Shaoshan.

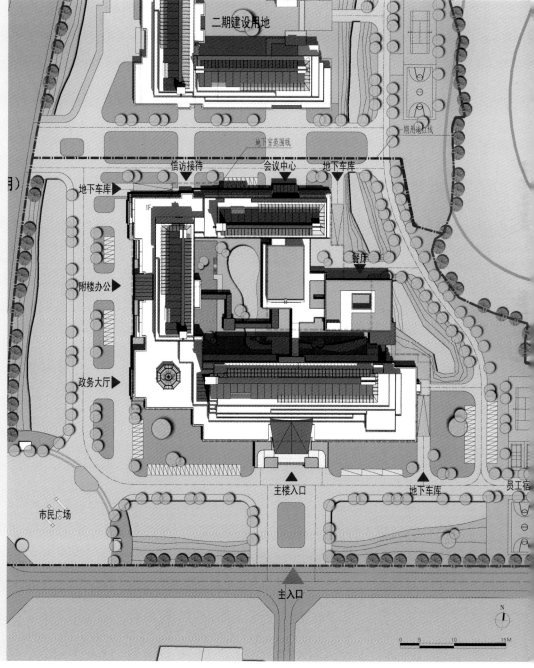

总平面 \ Site Plan

东南方向全景透视（陈 畅 摄影）\ Southeast Perspective (Photographer: Chen Chang)

建筑外部空间的处理主要围绕建筑与山水关系展开，东侧的虎形山，以及西南向的低洼湿地与建筑的有机组合，既提升了环境体量，也强化了建筑的开放性。

Design for architecture external space focuses on the relationship between architecture and environment. The Huxing Mountain on the east combines organically with the low-lying wetlands and building on the southwest, both enhancing the environment volume and strengthening the openness of the building.

主体建筑透视（陈 畅 摄影）\ Main Building Perspective (Photographer, Chen Chang)

局部（陈 畅 摄影）\ Local (Photographer, Chen Chang)

西南方向透视（陈 畅 摄影）\ Southwest Perspective (Photographer, Chen Chang)

建筑整体布局呈U字型，一幢主楼和两幢附楼形成一个朝虎形山打开的"三合院"。"三合院"的空间容纳了服务交流性的大型会议室、餐厅、健身休息等功能，这些功能由景观性的廊子亭榭和水景有机的串连为一体，既满足了办公楼的使用要求，又使得院内空间富有园林的空间趣味与意境。

The building is in overall layout of U-shaped, so that one main building and two ancillary buildings form a "three-section compound" opened toward Huxing Mountain. The "three-section compound" space accommodates the large conference rooms, restaurant, and fitness lounge. These functions are connected organically as a whole by the landscape porch pavilions and waterscapes, both meeting the using requirements of the office building, and adding the compound interior spaces with spatial interest and artistic conception of garden.

庭院空间 (陈 畅 摄影) \ Courtyard (Photographer, Chen Chang)

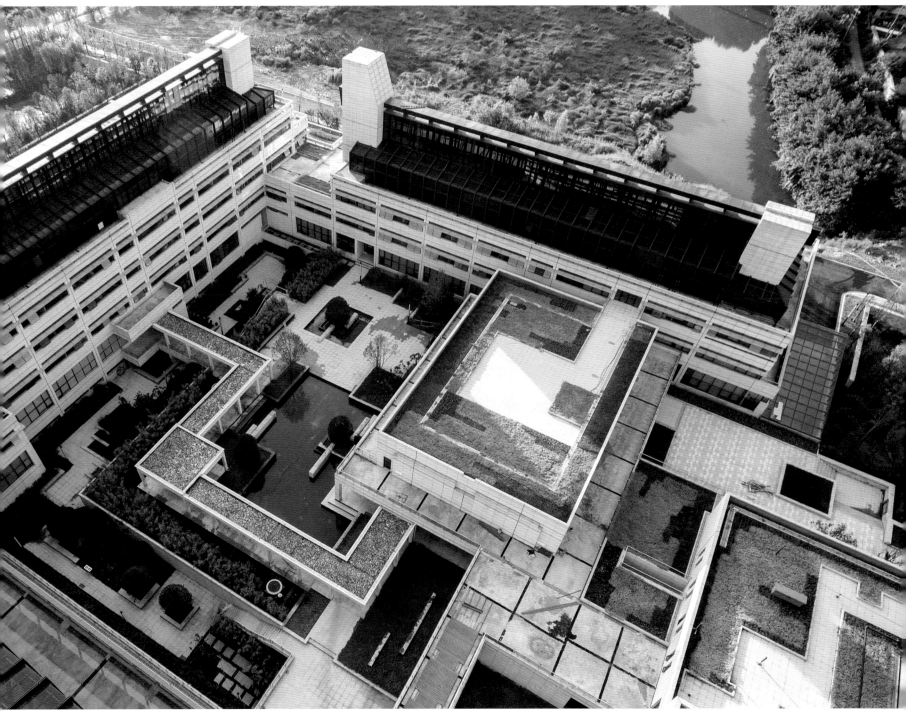

庭院鸟瞰（陈 畅 摄影）\ Aerial View of Courtyard (Photographer, Chen Chang)

庭院空间(陈 畅 摄影) \ Courtyard (Photographer, Chen Chang)

庭院内景（陈畅 摄影）\ Inner Garden View (Photographer, Chen Chang)

1、3 中庭空间内景（陈 畅 摄影）\ View of Atrium Space (Photographer, Chen Chang)
2 中庭二层内景（陈 畅 摄影）\ Inner View of Atrium Space at Second Floor (Photographer, Chen Chang)

杭州城市芯宇住宅小区
HANGZHOU METROPOLIS XINYU RESIDENTIAL AREA

合作者 鲁 华、徐 雄、陈 玲、吴妮娜、杨振宇、田 威、段继宗
设 计 2007 / 竣 工 2012

Co-designers: Lu Hua, Xu Xiong, Chen Ling, Wu Nina, Yang Zhenyu, Tian Wei, Duan Jizong
Design Time: 2007 / Completion Time: 2012

重新从城市的角度来审视居住建筑的设计模式
A design method of Re-examine residential buildings in an urban perspective

东北角透视（赵伟伟 摄影）| Northeast Perspective (Photographer: Zhao Weiwei)

设计出发点在于：重新从城市的角度来审视居住建筑的设计模式。在布局上，采用五座呈扇形排比的板式高层，与周围的建筑一道形成整体的城市空间形态。在造型上，结合日照要求进行了削角退台处理，并采用了大面积玻璃与铝板的对比，构成了本建筑独特的形态特征。小区内部则利用建筑底层5.6m的架空层，将绿化、小品引入其中，形成一个整体的园林景观。在住宅建筑技术上，本项目以住宅性能3A与绿色建筑3星为蓝本，采用了22项节能、减耗、智能化以及提高舒适度的设计系统，荣获中房协住宅最高奖广厦奖，并成为首批一星级绿建项目。

The design started from A Design Method of re-examine Residential Buildings in an Urban Perspective. In layout, 5 slab-block high-rises spread out in a sector, forming an overall urban spatial arrangement together with peripheral buildings. In configuration, beveled and retired desktop design in accordance with sun exposure requirement, and the contrast of big area glass and aluminum plate constitute the unique forms of the property. As to the internal residential area, the ground floors feature a 5.6m-high pilotis design, within which landscape features and garden have been introduced, forming an integrated garden-style landscape. In residential construction techniques, this project, using residential performance 3A and green 3-star as the prototype, has adopted a design system with 22 energy-saving, consumption-reducing, intellectualization, and comfort-improving indexes, It has won the Guangsha Award, the top residential prize, issued by the China Real Estate Association and become the first one-star green building project.

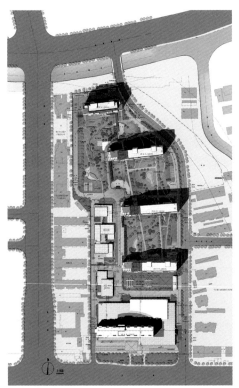

总平面 \ Site Plan

总体透视视(赵伟伟 摄影)\ Perspective (Photographer: Zhao Weiwei)

局部透视（陈 畅 摄影）\ Local Perspective (Photographer, Chen Chang)

东北角透视(赵伟伟 摄影) \ Northeast Perspective (Photographer, Zhao Weiwei)

建筑立面(赵伟伟 摄影) \ Facade (Photographer: Zhao Weiwei)

庭院景观（赵伟伟　摄影）\ Courtyard (Photographer：Zhao Weiwei)

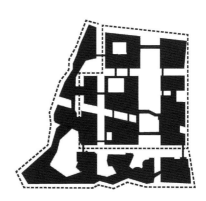

设计核心诉求是打造充满低碳理念和复合活力的顶级金融城,针对杭州自身的城市肌理以及本地块江河融汇的特征,提出"内城外郭"的规划结构,并且沿江河对角线的空间虚轴与内城的两条实轴交相辉映,呈现出"三轴一带"的空间格局,这样既彰显了钱江大潮之势,又体现了运河文化之韵。

Core design appeal is to create a top-class financial town with low-carbon notion and full of combined vitality. A structure of "town inside and wall outside" is proposed in accordance with Hangzhou specific municipal texture and local feature of river convergence. A space layout of "three axes and one belt" comes into view where imaginary space axis along river diagonal and two real axes enhance each other. This not only highlights momentum of Qiantang River Tide, but also represents charm of canal culture.

杭州钱江金融城概念方案
HANGZHOU QIANJIANG FINANCIAL AREA

合作者 薄宏涛、王大鹏、郑英玉、杨涛、刘翔华、柴敬、黄斌、王岳峰、李雯雯、戚卫娟、王政、桂汪洋、梁超凡、汪毅
合作单位 北京土人城市规划设计有限公司 艾奕康咨询（深圳）有限公司（AECOM）上海分公司
设计 2013 参加国际竞标入围

Co-designers: Bo Hongtao, Wang Dapeng, Zheng Yingyu, Yang Tao, Liu Xianghua, Chai Jing, Huang Bin, Wang Yuefeng, Li Wenwen, Qi Weijuan, Wang Zhen, Gui Wangyang, Liang Chaofan, Wang Yi
Cooperation Company: Turenscape, AECOM Shanghai
Design Time: 2013 International Competitive Bidding, Finalist

依江傍河展开的多层次空间
Multi-level space unfolded along river

沿江透视 \ Perspective along the Canal

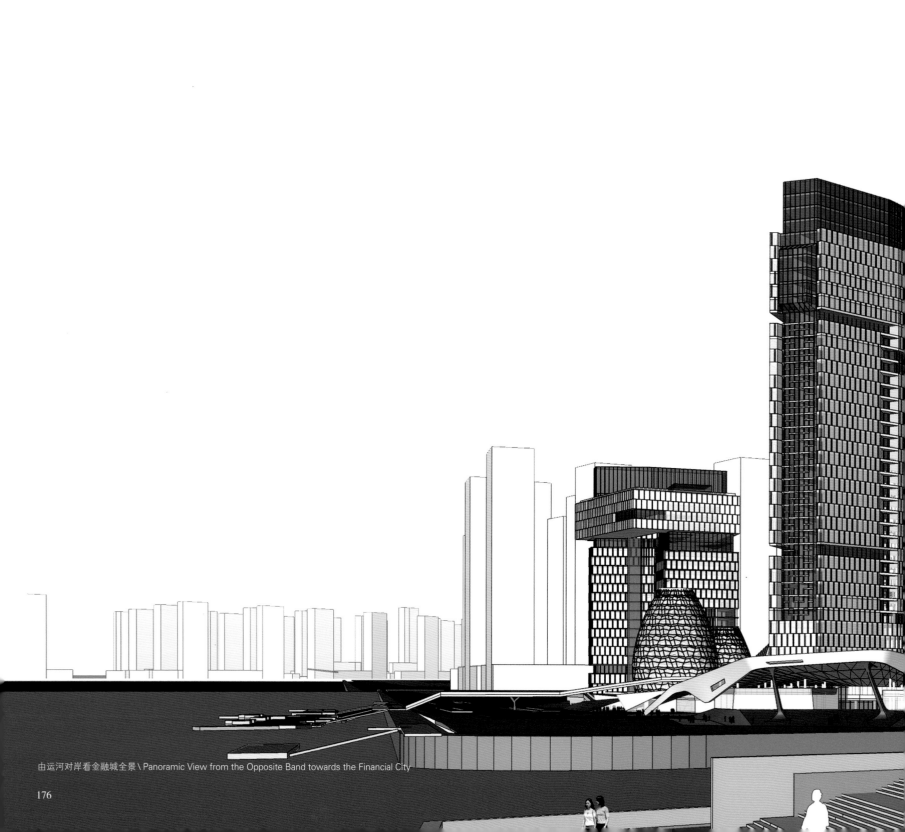

由运河对岸看金融城全景 \ Panoramic View from the Opposite Band towards the Financial City

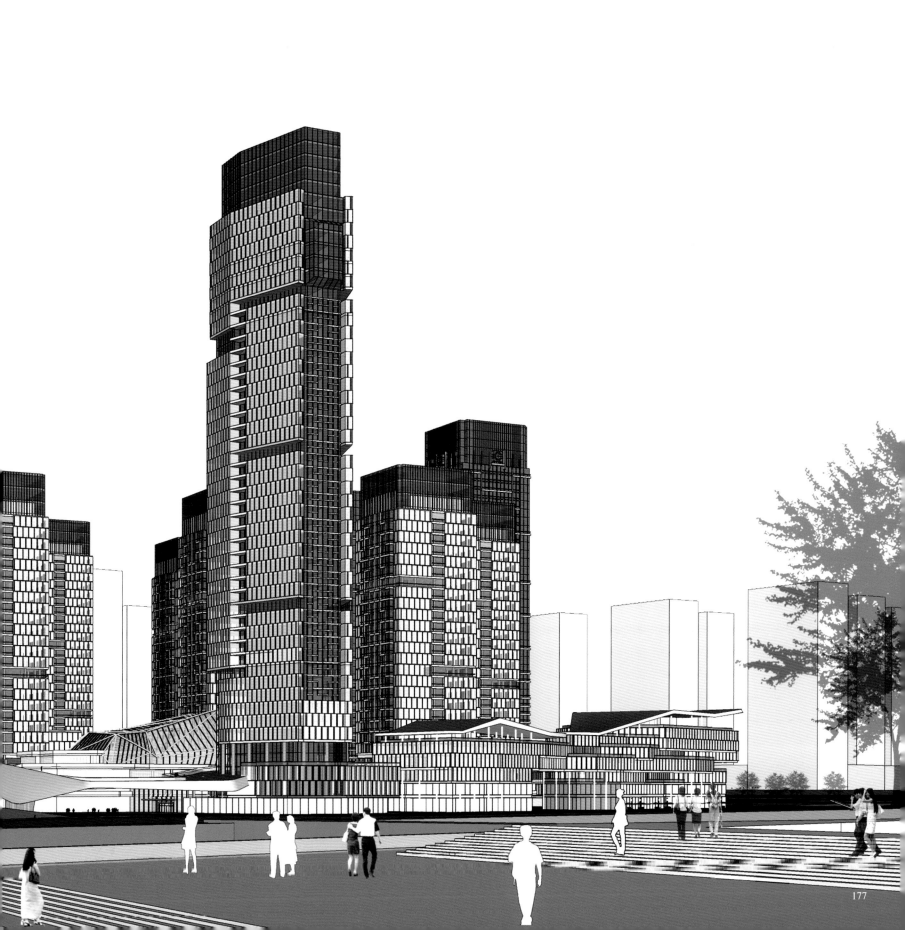

沿钱塘江透视 \ Perspective along Qiantang River

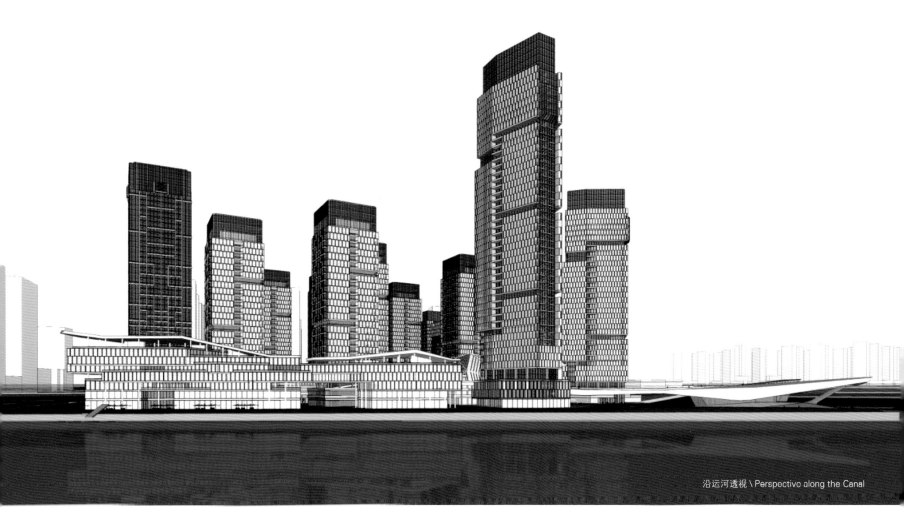

沿运河透视 \ Perspective along the Canal

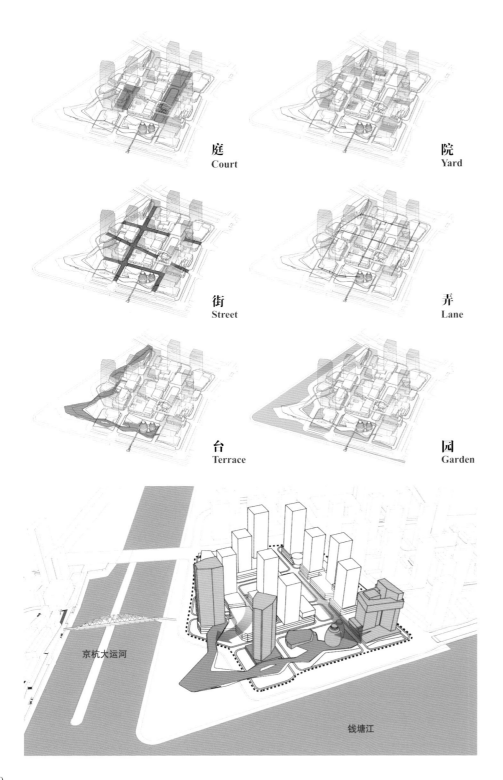

设计充分汲取中国传统空间营造经验，在场地内部打造出"庭"、"院"、"街"、"弄"、"台"、"园"六种不同性质的古典空间类型，且根据建筑类型及功能的不同，形成多样的空间形式，各空间之间建立明确、丰富且自然的过渡与转换方式。

The design fully absorbs experience in creation of Chinese traditional space to create six classic space types of different natures within the site, i.e., court, yard, street, lane, terrace, and garden. It further creates diversified space forms in accordance with different architectural types and functions with definite, abundant and natural transition and conversion between different types of space.

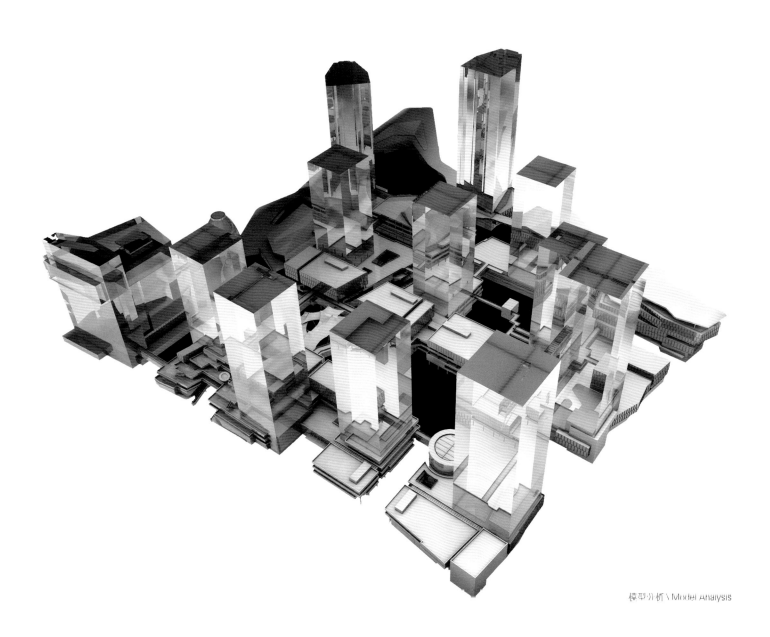

模型分析\Model Analysis

庭
Court

设计利用上位规划中原有中央景观广场形成主庭来容纳商业、景观及其他各种活动。

The main court is formed on the basis of the original central space to create a place for business,landspace and other activities.

院
Yard

设计利用塔楼间的拓扑关系及每个地块的裙楼设置多个庭院，形成疏密得当、有节奏的中国古典式庭院组合。

Several yards are formed among topological towers and lower blocks to make appropriate and rhythmic classical combinations of yards in Chinese-style.

街
Street

设计保留上位规划的街道设置，并利用主庭将其有效整合在一起，闻涛路一线打造步行商业街，并利用两侧商业形成富有活力的城市公共活动空间。

All the streets in previous plan had been retained and integrated by main court in central area, and Wentao Road is transformed into an energetic pedestrian street and bustling public commercial spaces .

弄
Lane

结合杭州地区公共生活特征，在地块内部设置纵横交错的里弄体系，为地块增加另一层级的步行系统，使得地块之间及内部的空间联系更加高效和富有趣味性，并借助天井与水院营造出富有杭州传统特征的公共活动区域。

Considering Hangzhou traditional lifestyle, lanes are introduced to the public communication of the site to establish a more efficient and interesting pedestrian system, which can create a kind of native neighborhood with patios and water yards simultaneously.

台
Terrace

经二路沿线利用四层平台形成空中副庭，主副两庭负责统合、联系地块中其他各种形成的空间。

Along Jing Er Road, a fourth-floor platform made a sky-court. These two conrts connect other spaces of the site.

园
Garden

沿大运河及钱塘江一线打造园林式景观，并在两江交汇出形成主要滨水活动空间，城市生活与自然景观在这里完美地融合为一体。

Gardens are designed along Grand Canal and Qiantang river to form vivid waterfront spaces, in which city-life and natural scenes merge harmoniously.

沿江标志性塔楼 \ Tower Building along River

中央商业大道透视 \ Perspective of Centrial Commercial Avenue

江的剧场 \ River Theatre

滨江活力生态公园 \ Riverside Vital Eco-Park

地下商业 \ Underground Commercial Area

架空酒吧 \ Overhead Bar

湿地公园 \ Wetland Park

西安大明宫遗址博物馆概念方案
CONCEPTUAL DESIGN OF DAMING PALACE SITE MUSEUM, XI'AN

合作者 薄宏涛、刘鹏飞、唐 斌、于 晨、蒋 珂、单晓宇、应 瑛、杨 涛、徐勤力
设 计 2009 未实施

Co-designers: Bo Hongtao, Liu Pengfei,Tang Bin,Yu Chen, Jiang Ke, Shan Xiaoyu,Ying yin, Yang Tao, Xu Qinli
Design Time: 2009 Unimplemented

出土的『城』
Image of Unearthed "Town"

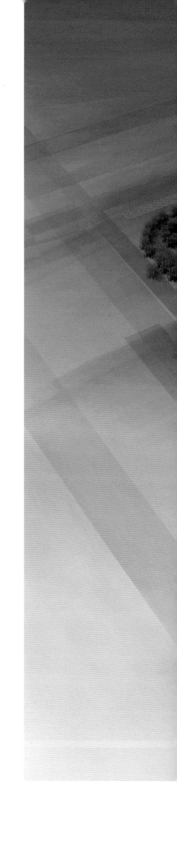

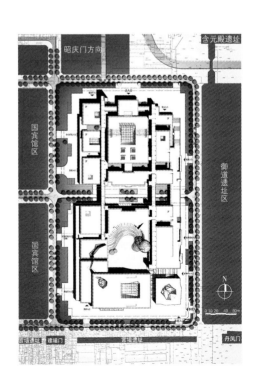

压低建筑高度，形成台地景观，力求避免对北面含元殿的影响 \ The Depressed building height forms a terrace scene and avoids influence on Hanyuan Hall to the north.

自含元殿基台高度半俯瞰建筑群 \ Perspective of the Building at Roof Height

科技馆主公共室外台阶向丹凤门方向透视 \ Perspective from the Main Public Out-door Steps of Science and Technology Museum to Danfeng Gate

室外楼梯 \ Outdoor Stairs

项目位于大明宫国家遗址公园殿前区中轴线西侧,中轴线北面为含元殿遗址,是唐大明宫国家遗址公园的旅游配套及服务设施区。

在总平面布局上,我们将整块用地分成南北两大区块,将主要的博物馆建筑设于用地的北侧,科技馆和电影馆整合成一个整体区块设于用地的南侧。并通过两条贯穿南北区块的交通连廊将三馆有机地连为一体。合理分区,从而使各功能完备,设施先进,展藏研究与文化休闲一体化,提升整体环境品质。

为减少对大明宫遗址的影响,在符合功能要求的前提下,提高覆盖率,压低建筑高度,造型整体而质朴,宛如地景,营造一种出土的"城"的印象。

因考虑对含元殿遗址的影响,此方案未实施。

The project lies in the west of central axis of the Daming Palace National Heritage Park front area and in the north is the Hanyuan Hall site, and serves as the travel support and service facility area of Tang-Dynasty Daming Palace National Heritage Park.

For site plan, we divide the land into northern block and southern block by arranging the principal museum buildings in the north of the land, and by combining the Science and Technology Museum and Film Pavilion into an integral block in the south of the land. The three pavilions are connected into a whole through two connecting corridors that penetrate the north and south blocks. Rational zoning is performed to assure all-around function and advanced facilities, and thereby achieving the integration of collection research and leisure & culture, and improving the overall environment quality.

In order to reduce the effect of the Daming Palace Ruins, increase the coverage ratio and reduce the building height while assuring the compliance with functional requirements, the appearance is designed in an integral and unadorned manner to be like a ground landscape that creates an image of unearthed "town".

Considering the influence to the Haiyuan Palace (key palace of Tang-Dynasty Palace), this conceptual design is unimplemented.

博物馆主入口透视 \ Perspective of Main Entrance

科技馆弧线展廊室内透视 \ Interior Perspective of Curved Exhibition Gallery of the Science and Technology Museum

博物馆架构通廊室内透视 \ Interior Perspective of the Framework Corridor of the Museum

博物馆主展厅室内透视 \ Interior Perspective of Main Exhibition Hall of the Museum

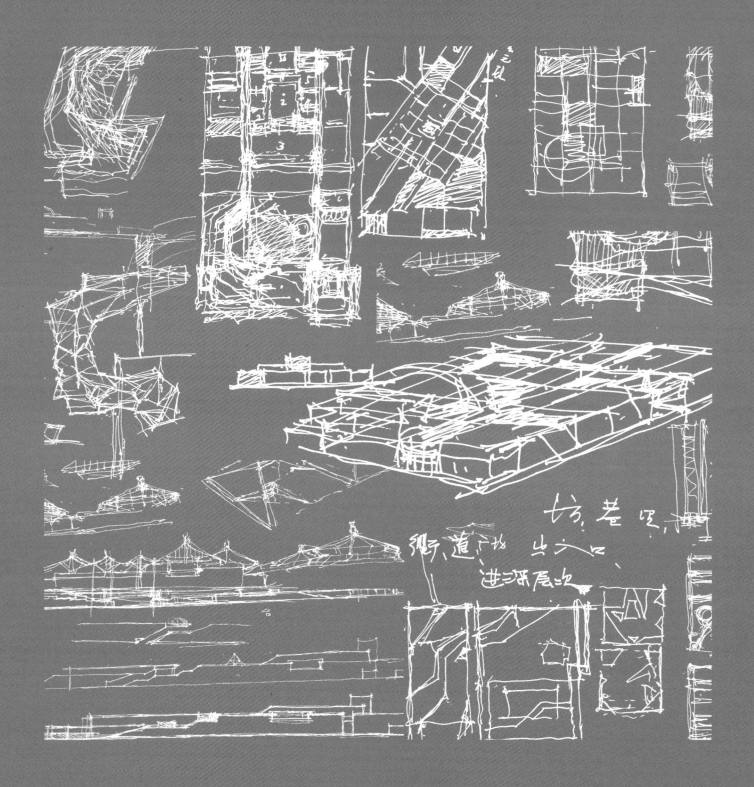

草图研究 \ Sketches

温岭博物馆
WEN LING MUSEUM

合作者 陈玲、刘翔华、王忠杰、史晟
设　计 2011 施工中

Co-designers: Chen ling,Liu Xianghua,
Wang Zhongjie, Shi Sheng
Design Time: 2011 Under Construction

石夫人山下的顽石
中国调性的非线性建筑造型

The stone at the foot of Ms. Shi Mountain
Non-Linear Architecture form with Chinese Aesthetics

横湖北路方向透视 \ Perspective towards North Henghu Road

博物馆位于规划的新城中心，石文化是浙江省温岭市主要的文化特征，新城又处在石夫人山山峦的怀抱之中，采用非线性的有中国特色的石头造型是一种十分自然的选择。让博物馆建筑与周围成片的高层建筑区分开来，以凸显其标志性建筑的地位，这是与规划要求相符的。

数字"语言"作为方法和手段，大大地拓展了建筑艺术的表现力。

The Museum is located in the center of the planned new city and the stone culture is a major cultural characteristic of Wenling City. The nonlinear stone shape with the Chinese features is a very natural choice. Separating the museum from the surrounding blocks of high-rise buildings in order to highlight it as the symbolic building in the city is in consistency with the planning. As the method and approach, the digital language greatly expanded the expression of the building arts.

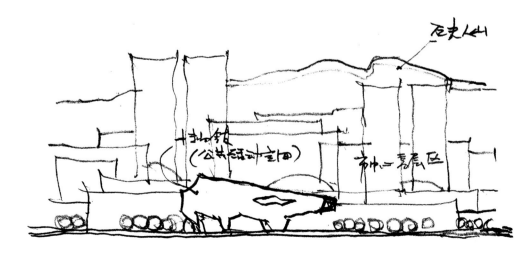

全景鸟瞰 \ Panoramic Aerial View

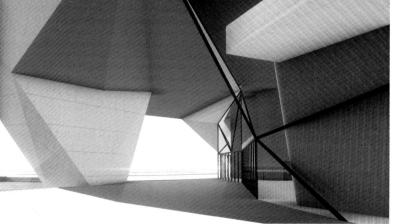

局部透视 \ Detailed Perspective

沿河透视 \ Along River Perspective

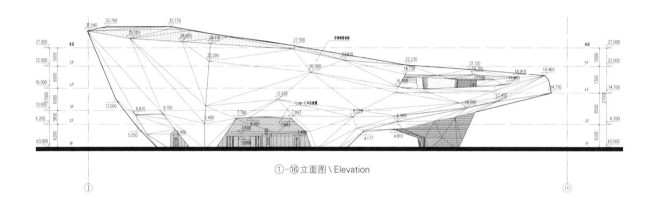

①-⑯ 立面图 \ Elevation

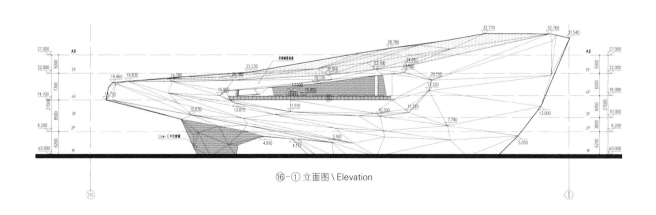

⑯-① 立面图 \ Elevation

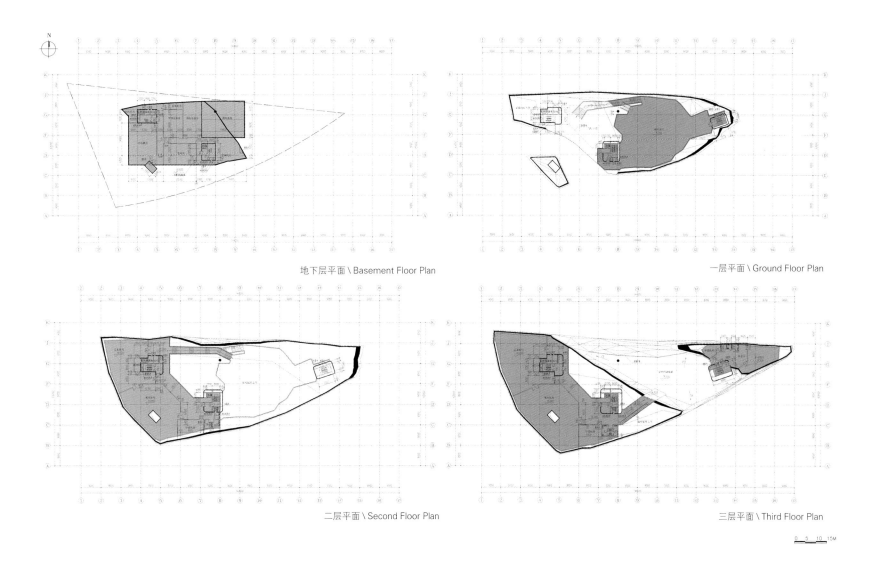

地下层平面 \ Basement Floor Plan

一层平面 \ Ground Floor Plan

二层平面 \ Second Floor Plan

三层平面 \ Third Floor Plan

室内透视 \ Interior Perspective

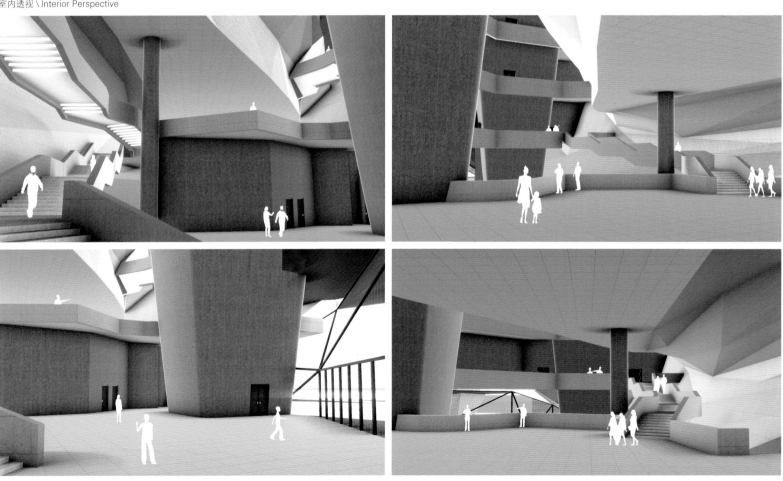

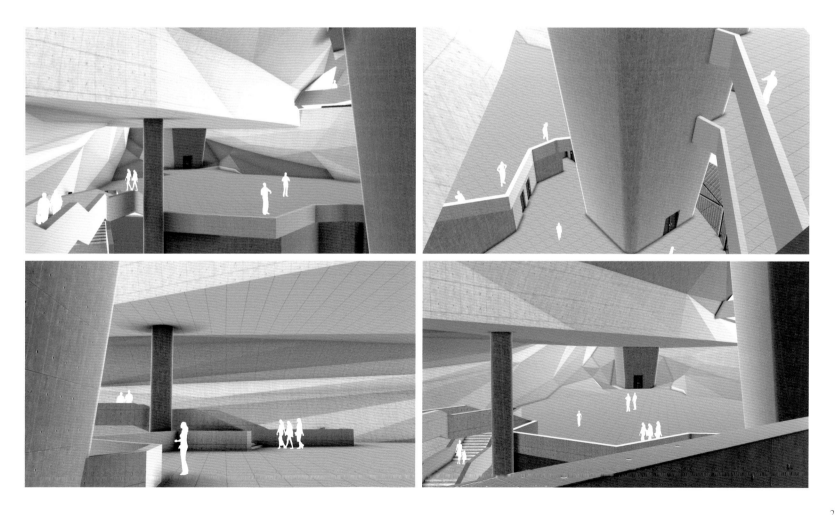

苏州越城遗址博物馆
SUZHOU YUE CITY SITE MUSEUM

合作者　殷建栋、钟承霞、朱文婧、桂汪洋、刘翔华
设　计　2013　设计中
Co-designers: Yin Jiandong,Zhong Chengxia,Zhu Wenjing,
Gui Wangyang,Liu Xianghua
Design Time: 2013　In Designing

「穴居式」博物馆
"Cave dwelling" Museum

全景鸟瞰 \ Panoramic Aerial View

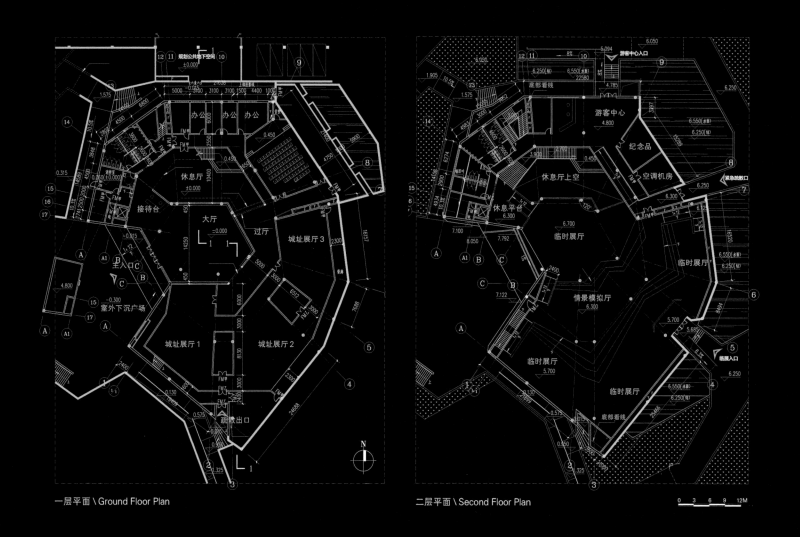

一层平面 \ Ground Floor Plan

二层平面 \ Second Floor Plan

建筑宛如"穴居"，参观者由地面甬道进入地下展厅，然后上至以"场景"为主的地上大厅。大厅屋顶由不规则的石板"搭"建，光线由石板缝隙射入。营造了一种沧桑、野趣的历史氛围。而且由此化解了过大体量，与周边渔家村建筑的体量取得协调。

The buildings are just like "living caves", where the visitors could enter underground exhibition hall through a corridor before going to the above-ground vitual exhibition of scenes. The hall roof is composed of irregularly shaped stone boards "stacked" in a certain manner, and the light passes through the joints between boards, and thereby creating a "weather-beaten" countryside-like historic atmosphere. Furthermore, the excessive large building volume is broken into pieces in this way to coordinate with the building volume of fishing villages around.

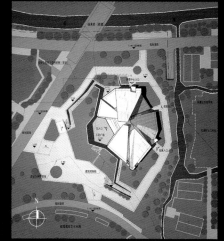

总平面 \ Site Plan

博物馆西侧透视 \ West Perspective

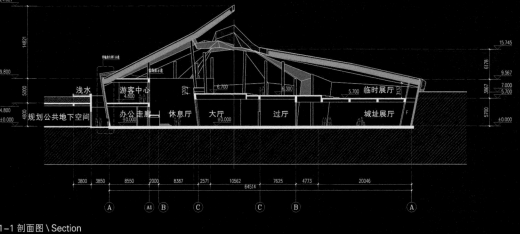

1-1 剖面图 \ Section

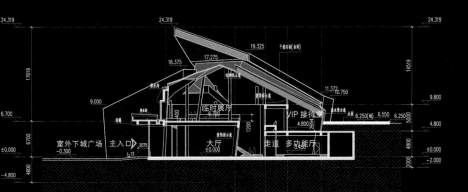

博物馆西南入口透视 \ Perspective of Southwest Entrance of the Museum

地面展厅内景（场景陈列）\ Interior View of Ground Exhibition Hall

由地下展厅上地面展厅 \ From Underground Exhibition Hall up to the Ground One

山西太原晋阳新城展示馆概念方案
CONCEPTUAL DESIGN OF JINYANG NEW TOWN EXHIBITION HALL, TAIYUAN, SHANXI

合作者 陈 玲、朱文婧、裘 昉、祝 容、汤 焱
设 计 2012 未实施
Co-designers: Chen Ling, Zhu Wenjing, Qiu Fang, Zhu Rong, Tang Yan
Design Time: 2012 Unimplemented

非线性语言的运用
青铜出水
Utilization of non-linear language,
Like bronze rising from water

总平面 \ Site Plan

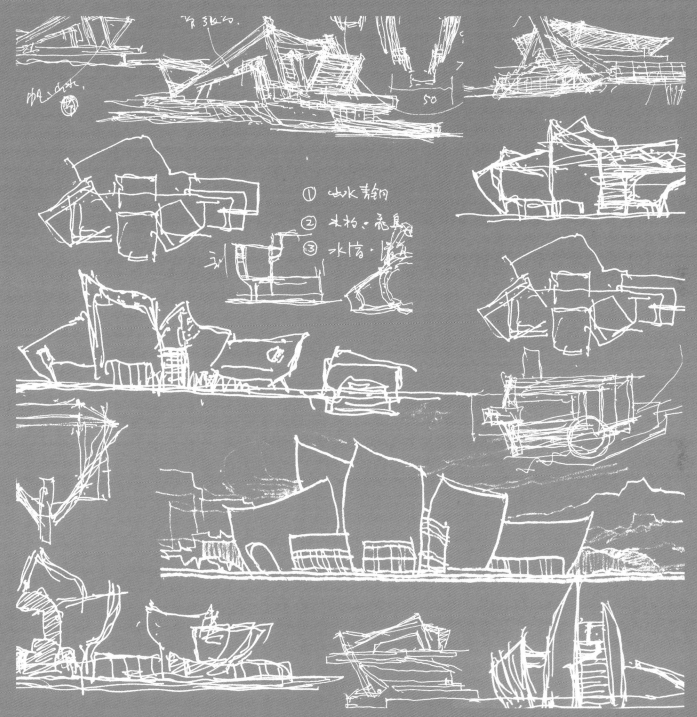

草图研究 \ Sketches

主体透视 \ Main Building Perspective

晋阳为春秋古城，青铜的制造和使用是该时期文化发展的重要标志。以"青铜出水"作为晋阳湖展示馆的创意，把历史、文化和场地环境有机结合起来。

展示馆包括：序言区、模型区、互动体验区、建设历史及规划展示区、管理办公区、接待商谈区、会议室、休闲区等多个功能区域，设计时把以上功能整合为各个单元，与中庭相联系。

Jinyang is an ancient city in Spring and Autumn Period. Creation and utilization of bronze is an important symbol of cultural development in this period. "Bronze rising from water" which serves as a creativity of Jinyang Lake Exhibition Hall has combined history, culture and onsite environment organically.

Exhibition building includes preface area, model area, interactive experience area, construction history and planned exhibition area, management office area, reception and negotiation area, meeting room and entertainment area. The above stated functions are integrated into different units and related with courtyard.

主体透视 \ Main Building Perspective

苏步青纪念馆
SU BUQING MEMORIAL HALL

合作者 陈 玲、刘辉瑜、王忠杰、朱文婧、张天钧、李 照、宋一鸣
设 计 2011 即将竣工
Co-designers: Chen Ling,Liu Huiyu,Wang Zhongjie,Zhu Wenjing,Zhang Tianjun,Li Zhao,Song Yiming
Design Time: 2011 Nearly Complete

几何构成,向大师致敬
Geometrical structure to pay homage to the Master

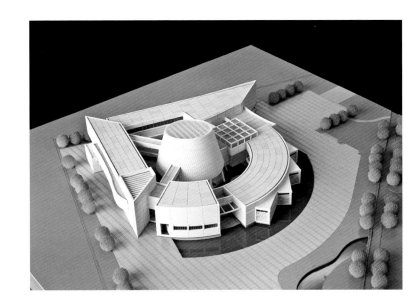

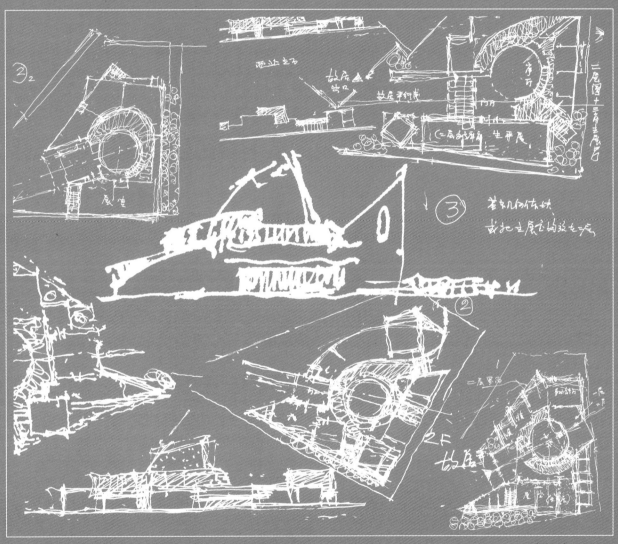

草图研究 \ Sketches

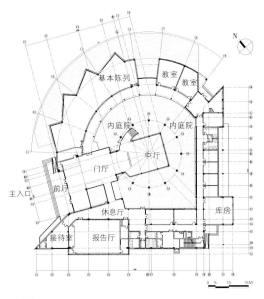

一层平面 \ Ground Floor Plan

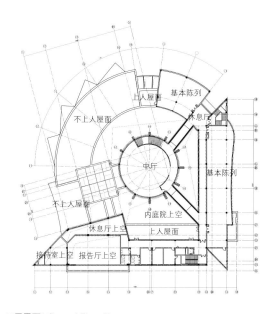

二层平面 \ Second Floor Plan

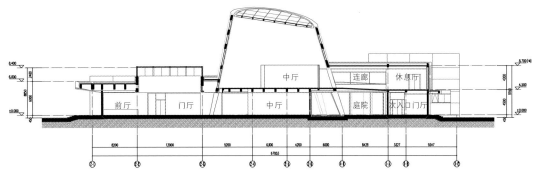

剖面图 \ Section

全景鸟瞰 \ Panoramic Aerial View

由于整个区域以低层建筑为主，因此基地内建筑设计为1-2层。靠近故居与亭子街的方向建筑高度相对较低，为1层高，东侧、南侧建筑相对较高。在中心的圆锥形简体为建筑的最高点，建筑错落有致。

本方案从数个基本几何形体出发，隐喻了苏步青先生的著名微分几何数学大师身份；同时通过对几何形体的变形对周边环境和故居进行对应和协调。

建筑空间布局，本案以圆锥形大厅为核心，沿周边布置庭院与展览空间，形成虚实相间、富于层次变化的建筑空间。

Since the whole area is dominated by low buildings, this building in the site has 1~2 floors. Buildings near former residence and pavilion street have lower height, i.e. Buildings on the east side and the south side are relatively higher. Cone-shaped structure in the center is the apex of the buildings which are well arranged.

This scheme originates from several basic geometric forms to symbolize identity of Mr. Su Buqing as a famous differential geometry master. It also corresponds and coordinates surrounding environment and former residence through geometric deformation.

Regarding to the architectural space layout, this scheme arranges courtyard and exhibition space around cone-shaped hall to form architectural space of virtual and real and rich hierarchical variation.

东南角透视图 \ Southeast Corner Perspective

正立面透视图 \ Perspective of Main Entrance

庭院透视 \ Perspective of Courtyard

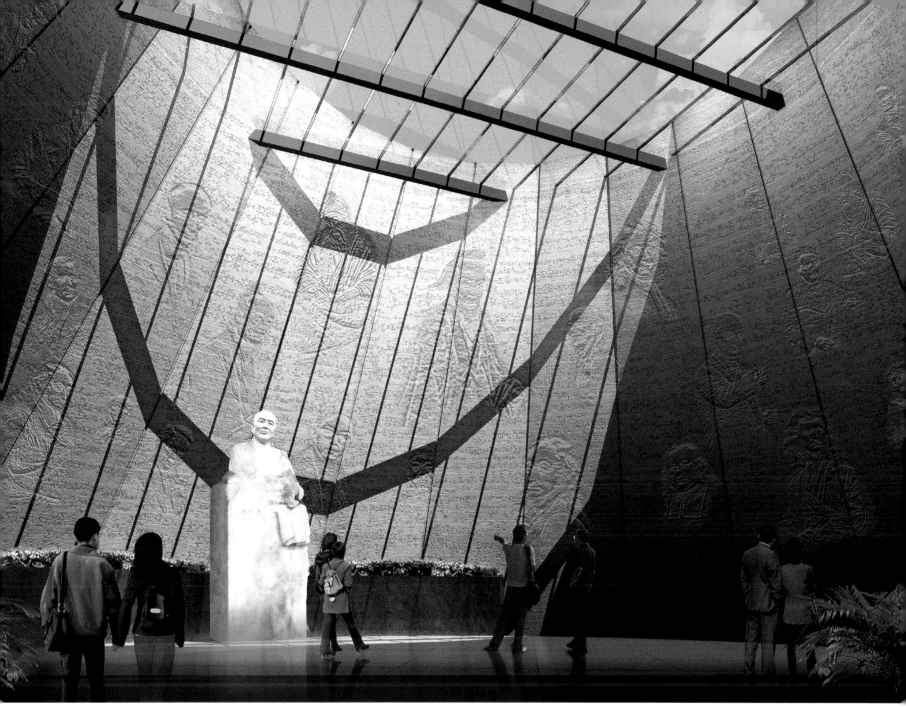

中央展厅透视 \ Perspective of Central Exhibition Hall

杭州师范大学仓前校区B组团
CANGQIAN CAMPUS (B BLOCK) OF HANGZHOU NORMAL UNIVERSITY

湿地书院
Wetland Academy

合作者 王大鹏、柴 敬、周炎鑫、庄允峰、谢 悦、沈一凡、黄 斌、王岳锋、陶 涛、汤 焱、祝 容、胡晓明、贾秀颖

设 计 2013 施工中

Co-designers: Wang Dapeng,Chai Jing,Zhou Yanxing,Zhuang Yunfeng, Xie Yue,Shen Yifan,Huang Bin,Wang Yuefeng,Tao Tao,Tang Yan, Zhu Rong,Hu Xiaoming,Jia Xiuying

Design Time: 2013　Under Construction

生活区中心透视 \ Perspective of Residential Area

01	新能源学院 \ New Energy School	13	工科实验楼一 \ Engineering Course Experiment Building I
02	材化学院 \ Material Chemistry School	14	工科实验楼二 \ Engineering Course Experiment Building II
03	公共实验楼一 \ Public Experiment Building I	15	专家公寓 \ Expert Apartment
04	公共实验楼二 \ Public Experiment Building II	16	教师公寓一 \ Teacher's Apartment I
05	数理类教学楼 \ Teaching Building for Mathematics And Physics	17	教师公寓二 \ Teacher's Apartment II
06	军工类实验楼 \ Experiment Building for Military Industry	18	本科生公寓一 \ Undergraduate Student Apartment I
07	遥感技术学院 \ Remote Sensing Technology School	19	本科生公寓二 \ Undergraduate Student Apartment II
08	计算机学院 \ Computer School	20	研究生公寓 \ Graduate Student Apartment
09	国际服务外包学院 \ International Service Sourcing School	21	学生活动中心 \ Student Activity Center
10	公共教学楼二 \ Public Teaching Building II	22	物业管理用房 \ Property Management Room
11	图书馆分馆 \ Library Branch	23	食堂 \ Dining Hall
12	公共教学楼一 \ Public Teaching Building I	24	产学研综合体 \ Industry-University-Research Complex

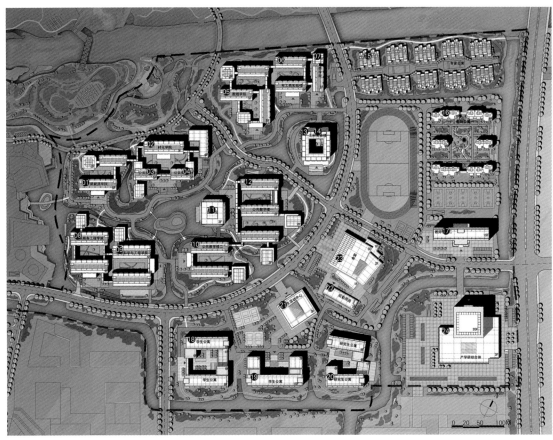

总平面 \ Site Pan

教学区鸟瞰 \ Aerial View of Teaching Area

1	3
2	4

1 底层架空层空间 \ Open Floor Space at Grourd Floor
2 中轴线架空层空间 \ Open Floor Space at Axis Zone
3、4 架空连廊空间 \ Overhead Corridor Space

整体设计遵循"湿地书院"与"和而不同"的理念，构思充分汲取湿地肌理、传统书院的精华，其因地制宜、情景交融的规划布局，疏密有致的院落簇群与融传统与现代于一体的建筑单体，泛中心感的形态和适度复合的功能，创造了具有杭师大文化传承的开放式、生态化、园林式校园空间。建筑造型的构思主要来源于对中国传统文字构架和杭州历史建筑的认知，梁柱体系、坡屋顶、窗扇花格等原型要素的转换重现，形式在"似与不似之间"，既体现了灵动、雅致的地域特点，又具大气简约的时代性。

The overall design complies with a notion of "wetland academy" and "harmony in diversity", which fully absorbs essence of wetland texture and traditional academy. Its planned layout is suitable to local condition and integrated into scenes. Properly arranged courtyard cluster, single building which integrating tradition and modern, multi-centre-like form and properly combined function create open, ecological and garden-style campus space inheriting traditional culture of Hangzhou Normal University. The form of the building originates from cognition of Chinese traditional text structure and Hangzhou's historical building. Conversion and recurrence of post/beam system, sloping roof, window lane and other prototype elements as well as form between similarity and dissimilarity reflect regional features of flexibility and elegance as well as modern feature of liberality and simplicity.

材化学院东侧透视 \ Perspective of East Side of Material Chemistry School

厦门悦海湾酒店
YUEHAI BAY HOTEL XIAMEN

合作者 殷建栋、吴妮娜、杨 涛、庄允峰、闵 杰、周 逸、朱文婧、裘 昉、袁 越、陈 鑫、刘翔华、曾德鑫、郑建国、郑克卿、周 慧
合作设计单位 中建东北院厦门分院
设 计 2012 设计中

Co-designers: Yin Jiandong,Wu Nina,Yang Tao,Zhuang Yunfeng,Min Jie,Zhou Yi,Zhu Wenjing,Qiu Fang,Yuan Yue,Chen Xin,Liu Xianghua,Zeng Dexin,Zheng Jianguo,Zheng Keqing,Zhou Hui
Cooperation Design Company: Xiamen Branch of Northeast Institution of CSCEC
Design Time: 2012 In Designing

与大海、城市空间对话后的建筑生成
The architecture is generated from the "dialogue" between the sea and urban space

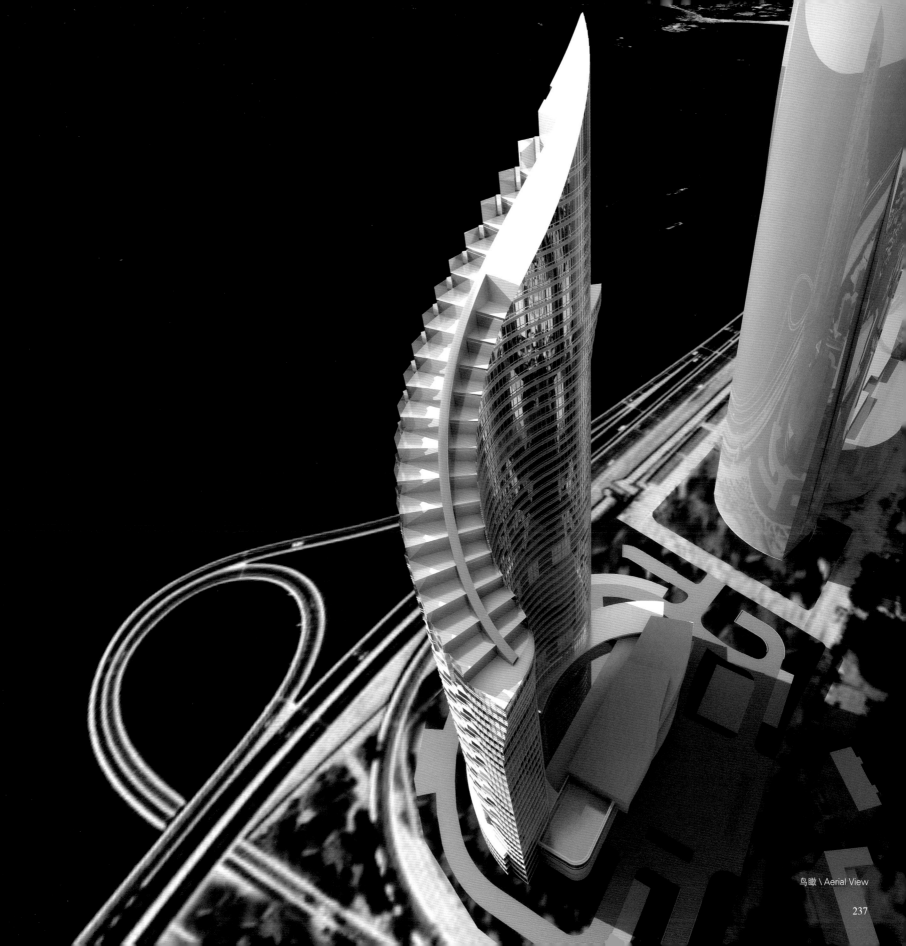

鸟瞰 \ Aerial View

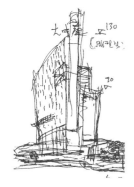 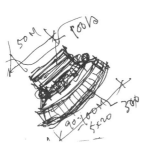

弧板
全海景客房　流线造型　连续空间　高度优势

Cambered Surface Tower
Guest Room with Full View of Sea, Streamline Modeling,
Continuous Space, Altitude Advantage

 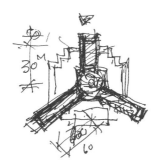

三板
大界面海　修长挺拔　古典美学　中式意象
高识别度

Three Individual Towers
Large-interface Sea, Tall and Straight, Classical Aestheticism,
Chinese Imagery, High Resolution

 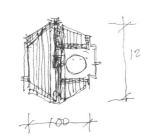

双板
动感造型　独特中庭　丰富空间　高识别度

Two Connected Towers
Dynamic Modeling, Unique Courtyard, Abundant Space,
Highly Identify

 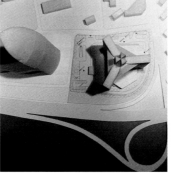 

方案比较 \ Proposals

本案采用"鹭起鸥落，云卷云舒，鳞次栉比，相依相携，天人无二，不必言"的设计理念，意在打造一个更自然、更城市、更中国的现代酒店。我们选取多个比较有代表性的场地角度作为设计的切入点，对每个角度经过慎重的考虑，以很好地回应场地的需求。

This case employs the design concept described as "the gathering of aigrette and gull, the swirling clouds, row-upon-row arrangement, dependency & mutual assistance, and the harmony between human and nature" so as to build a more nature-friendly and more urbanized modern hotel with more distinctive Chinese characteristics.
We selected several typical angles as entry point of design and gave deliberate consideration to each angle in order to meet the need of location.

区位分析 \ Location Analysis

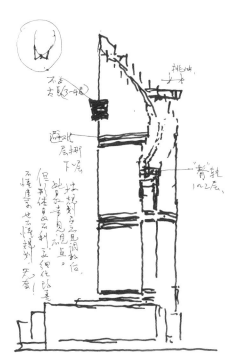

草图研究 \ Sketches

城市形象：在方案的开始，我们调研了周边的主要的城市交通和人流车流方向，研究建筑主要的形象视点，使其在城市中具有标志性，并与相邻的 300m 高双塔具有和谐的过渡关系。全海景房：塔楼采用单廊板式的结构，客房全部朝向海面，获得最大的城市展示面和一线海景。屋顶绿化系统：在裙房部分，我们采用将塔楼局部架空的做法来实现屋顶绿化朝向海面的视线通透性，同时结合裙房四层的泳池等休闲空间营造出惬意的景观空间。塔楼的顶部采用逐层退台的形式，结合独特的屋顶景观资源，打造一系列富有特色的高档海景客房。

CITY IMAGE: We first surveyed and studied the prevailing urban transportation and traffic flows in surrounding areas, and analyzed the main image viewpoint of building so that it could be developed into a landmark of the city and it's possible to establish a harmonious transitional relation with the adjacent 300m-high twin towers.

ROOMS WITH FULL VIEW of SEA: The tower is designed with a single-slab structure, where all the guest rooms face towards the sea so as to maximize the city display surface and the frontline seascape.

ROOF GREENING SYSTEM: Roof greening is realized at annex through "local overhead" of tower; the visual penetration towards the sea together with the swimming pool on the fourth floor of annex and other leisure spaces creates an agreeable landscape space. A series of full-drills top-grade sea view rooms are developed on the top of tower through floor-by-floor set-backmodel in conjunction with the unique roof landscape resources.

1 南海面东南向西北建筑形象 \ View from the Sea
2 鼓浪屿方向建筑形象 \ View from Gulangyu Island
3 鸟瞰 \ Aerial View

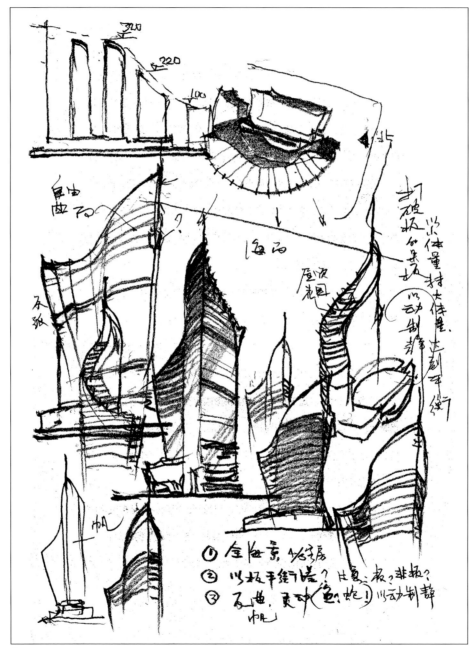

草图研究 \ Sketches

鸟瞰 \ Aeiral View

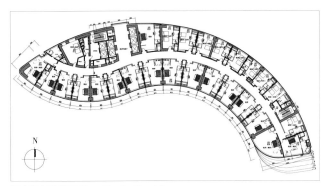
酒店客房层平面 \ Floor Plan for Hotel Rooms

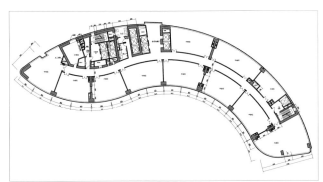
办公标准层平面 \ Standard Floor Plan for Offices

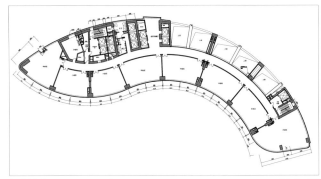
架空层平面 \ Open Floor Plan

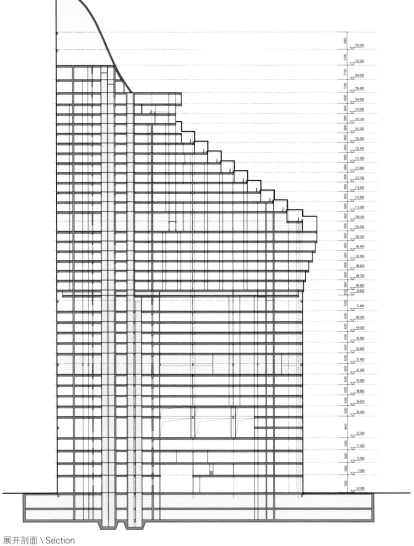
展开剖面 \ Section

环岛路透视 \ Perspective from Huandao Road

局部 \ Detailed Perspective

环岛路透视 \ Perspective from Huandao Road

上海杉杉控股总部大楼
SHANGHAI SHANSHAN GROUP HEADQUATERS BUILDING

合作者 殷建栋、杨涛、刘翔华、朱文婧、袁越、陈鑫、王政、古振强、周逸

设计 2013 设计中

Co-designers: Yin Jiandong, Yang Tao, Liu Xianghua, Zhu Wenjing, Yuan Yue, Chen Xing, Wang Zhen, Gu Zhengqiang, Zhou Yi

Design Time: 2013 In Designing

海上花
Flowers of Shanghai

夜景鸟瞰 \ Aerial View at Night

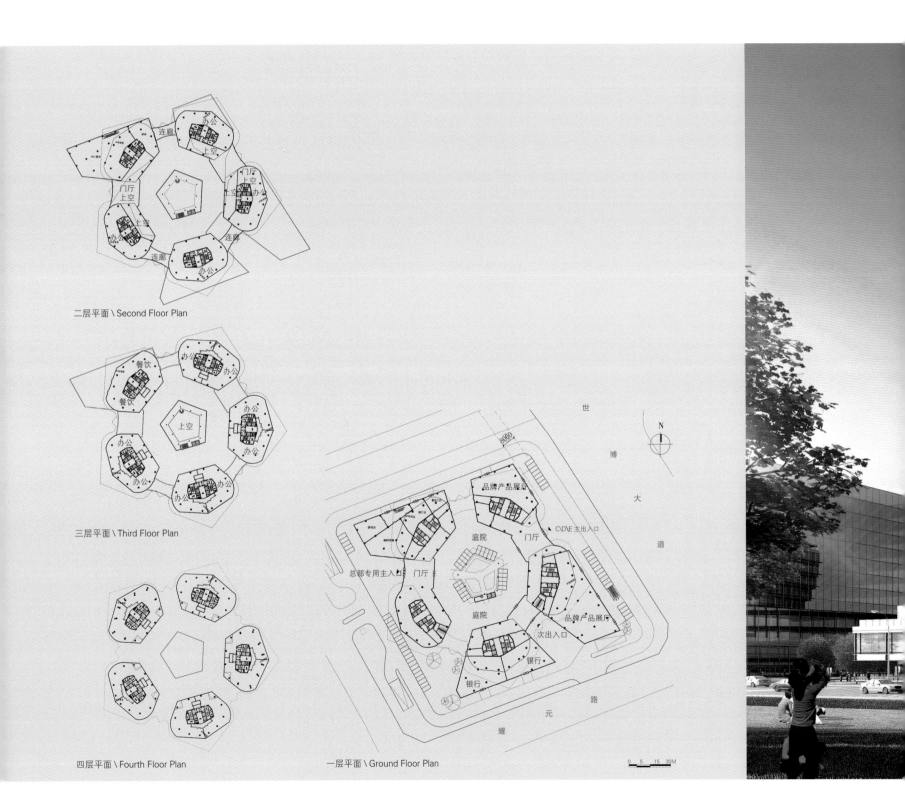

沿世博大道透视 \ Perspective along the Expo Avenue

本项目基地位于后滩CBD，世博园区南侧，为三大新兴板块汇聚之地，处于黄浦江滨江第一界面，西侧紧邻沿江绿化带，东侧为世博大道。设计以"海上花"的形体意象，来回应城市，凸显沿江地标形象，回应业主"杉杉之花在上海绽放"的愿景。同时，办公空间、景观空间、商业空间、城市空间的相互渗透交融，力求打造一个绿色生态的立体山水园林。

This project is located in Shanghai Houtan CBD and on the south side of Expo Park and is convergence of three emerging plots. It serves as the first interface of Huangpu River neighbors riverside green belt on the west and borders Expo Avenue on the east. Its design image of "Flower of Shanghai" corresponds to the city, highlights its landscape position along river and realizes the Employer's vision of "Shanshan's flower blooming in Shanghai". Meanwhile, office space, landscape space, commercial space and city space are mutually penetrated and integrated to form a green, ecological, stereoscopic landscape garden.

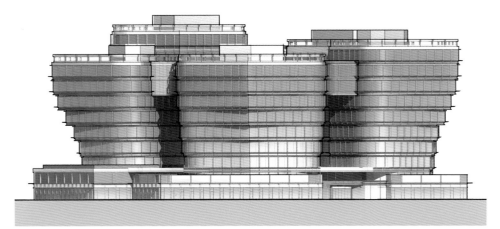

东立面 \ East Elevation

沿耀元路透视 \ Perspective along Yaoyuan Road

无锡锡东新城文化中心概念方案
CONCEPTUAL DESIGN OF XIDONG NEW TOWN CULTURAL CENTER, WUXI

合作者　殷建栋、朱文婧、郑克卿、董雍娴、周　慧、王碧君
设　计　2012

Co-designers: Yin Jiandong, Zhu Wenjing, Zheng Keqing, Dong Yongxian, Zhou Hui, Wang Bijun
Design Time: 2012

寓「丰」于「简」，寓「动」于「静」
Reflect abundance through simplicity and reflect activity through static.

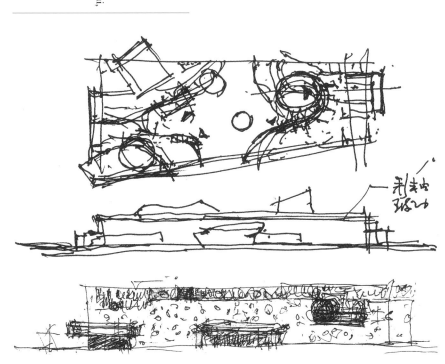

鸟瞰 \ Aerial View

该方案功能整合集约，造型简洁现代。临锡沪路形成连续生动的城市界面，同时也为北侧的广场提供了相对较为安静舒适的环境。建筑外立面采用穿孔铝板镂空图案表皮处理手法，建筑通透纯净，生动有变化。内部收放有致，空间具有趣味性。

This scheme provides integrated and intensive function with concise and modern form. A continuous and vivid municipal interface formed along Xihu Highway also provides relatively quiet and comfortable environment for a square on its north side. Exterior architectural façade is made of perforated aluminum plate with engraved pattern, demonstrating transparent purity and vivid variation. Its inside boasts well-arranged flexibility and interesting space.

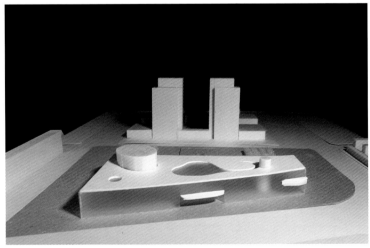

模型分析 \ Model Analysis

主体透视 \ Perspective of the Main Building

室内空间 \ Interior Space

室内空间 \ Interior Space

湛江文化艺术中心方案
PROPOSAL OF ZHANJIANG CULTURAL CENTER

合作者：王大鹏、金 坤、杨 涛、刘翔华、朱周胤、胡蓓蓓、沈一凡、孙 铭、吕思扬、刘鹤群
设 计 2014 设计中

Co-designers: Wang Dapeng,Jin Kun,Yang Tao,Liu xianghua,Zhu Zhouyin,Hu Beibei,Shen Yifan,Sun Ming,Lv Siyang,Liu Hequn
Design Time: 2014 In Designing

风动南海好扬帆
不一样的环境，生成不一样的建筑
Sailing in windy South China Sea
Unique environment creates unique building

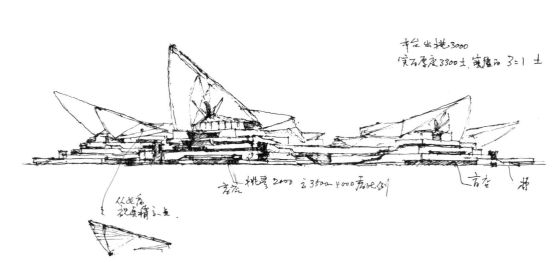

奥体中心方向夜景透视 \ Night Perspective towards the Olympic Sports Center

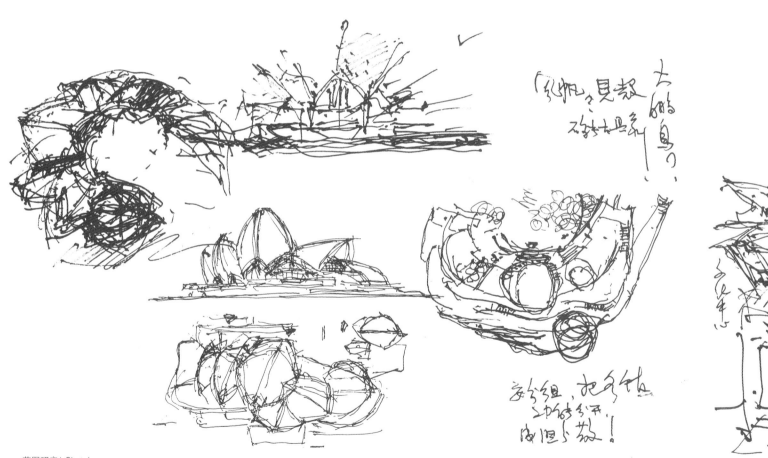

草图研究 \ Sketches

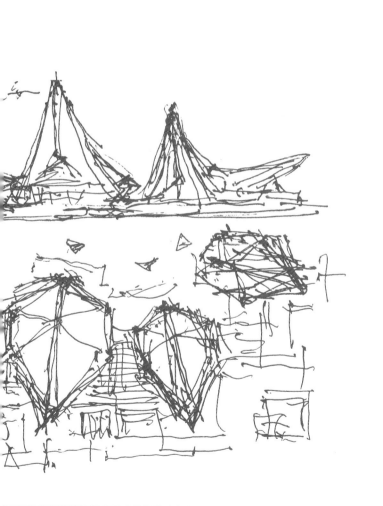
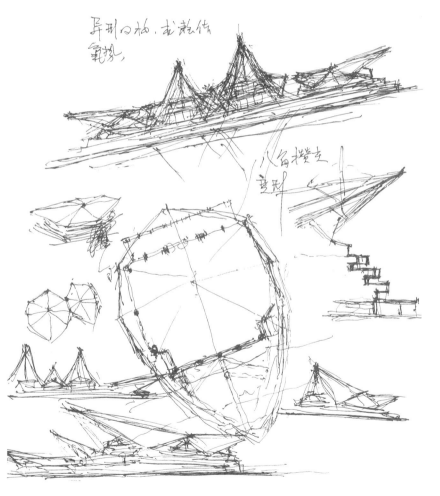

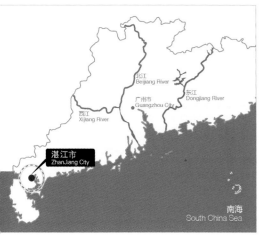

项目选址区域：广东省湛江市调顺岛军民大堤东南方向，本项目建设用地面积约 250 亩（含透水面积约 40 亩）。该地块地理位置独特，视野开阔，与美丽的金沙湾观海长廊隔水相望，毗邻正在建设中的滨湖公园，离湛江老城区距离较近，规划有一架桥梁，将其与赤坎区、海东新区相连接。

Project site: Located to the southeast of Army & People Dam on Tiaoshun Island of Zhanjiang City, Guangdong Province, this project occupies about 16.7 hectares land (including about 2.7 hectares of water permeation area). This site has a unique geological position, broad field of view, being separated from beautiful Golden Sand Bay sea-watching corridor across water, neighboring Binhu Park which is under construction and near to old town of Zhanjiang. It is planned to establish a bridge to connect it with Chikan District and Haidong New District.

基地位置 \ Site Location

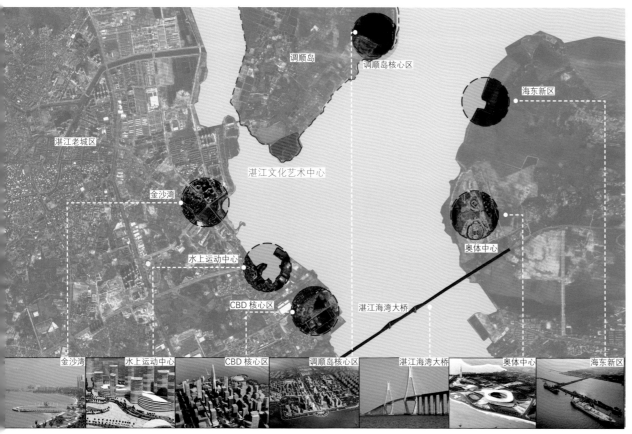

基地分析 \ Site Analysis

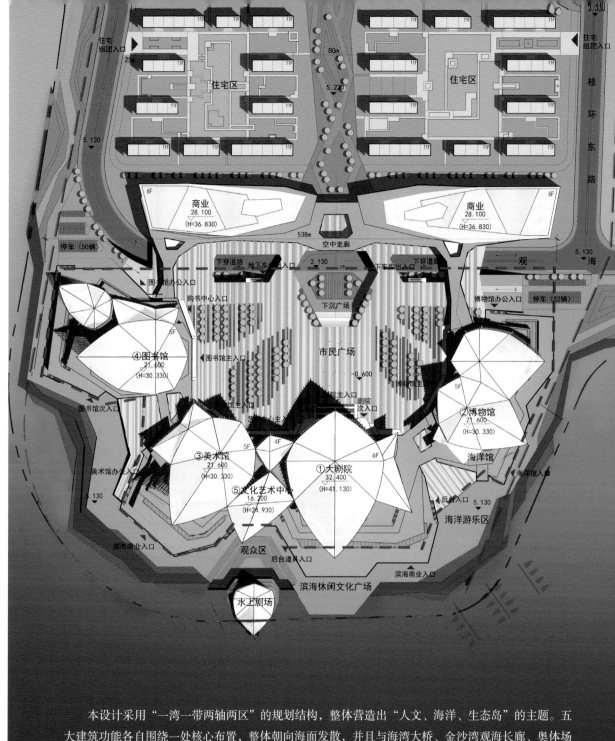

　　本设计采用"一湾一带两轴两区"的规划结构,整体营造出"人文、海洋、生态岛"的主题。五大建筑功能各自围绕一处核心布置,整体朝向海面发散,并且与海湾大桥、金沙湾观海长廊、奥体场馆、水上运动中心等分别形成相对位的视觉关系。各单体功能流线既相互独立又有联系,建筑公共空间连续开放,与滨海环境相衔接渗透,为市民营造出了良好的休闲文化活动场所。

This design adopts a structure of "one bay and one belt with two axes and two zones" to reflect the theme of "humanity, ocean and ecological island" as a whole. Five architectural programs are deployed around a kernel respectively. The overall layout diverges to sea surface and forms a visual relation of counterpoint with Bay Bridge, Jinsha Bay Watch Corridor, Olympic Sports Stadium and Water Sports Center. Each individual buildling is mutually independent and correlated. Public architectural space is continuously opened and is linked and penetrated into coastal environment to provide a good entertainment and cultural ocean for citizens.

总平面 \ Site Plan

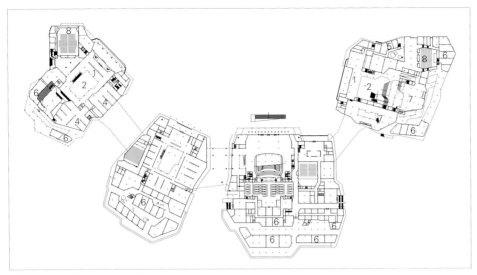

一层标高平面图 \ Ground Floor Plan

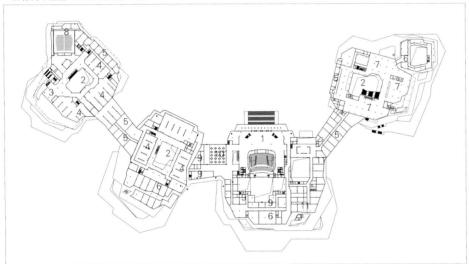

二层平面图 \ Second Floor Plan

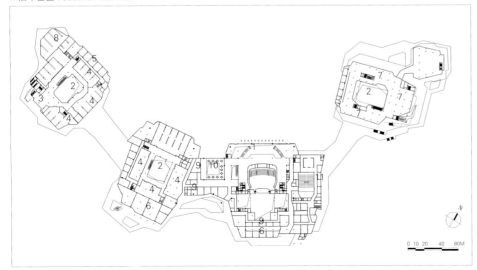

三层平面图 \ Third Floor Plan

1 门厅 \ Vestibule
2 中庭 \ Atrium
3 休息厅 \ Rest Room
4 展厅 \ Exhibition Hall
5 业务用房 \ Business Premises
6 商业 \ Commercial District
7 阅览室 \ Reading Room
8 报告厅 \ Lecture Hall
9 化妆间 \ Dressing Room
10 表演厅 \ Performance Hall
11 培训用房 \ Training Room

鸟瞰 \ Aerial View

夜景透视 \ Perspective at Night

海湾大桥方向透视 \ Perspective from the Bay Bridge

建筑形式明快而沉稳，精致而大气，典雅而豪迈，五大功能排列组合，形成连绵起伏、错落有致的以文化艺术为主题的生态滨海公园，文化中心与整体环境共同营造出闻涛起舞、舞动南海的意境。

Architectural form is sprightly and calm, delicate and liberal, elegant and magnificent. Permutated and combined five programs, the form is rolling and well-arranged biological Seashore Park with culture and art as theme. The cultural center and overall environment jointly create a prospect of dancing in accordance with ocean wave.

大剧院沿海面透视 \ Grand Theatre Perspective along the Coast

总体空间设计强调北侧市民广场与南侧滨海休闲文化广场的互动关系，两者通过建筑围合而成的灰空间相互交融、渗透，将滨海的活力和景观充分地引入到整个场地内。

建筑屋顶造型创意来自棕榈叶、折扇、竹伞、船帆等，它不仅让人可以联想到熟悉的生活场景，而且也具有遮阳挡雨的实际功能。

The overall space design stresses interaction between Citizen Square on the north side and Coastal Recreation & Culture Square on the south side. Through architectural enclosure, both squares bring out grey transitional spaces which integrate and penetrate each other to fully lead coastal vigor and scenery into the whole site.

Creativity of architectural roof modeling originates from palm leaf, folding fan, bamboo umbrella and boat sail. It not only reminds people of familiar scene in life but also plays a practical role of sunshade and rain shelter.

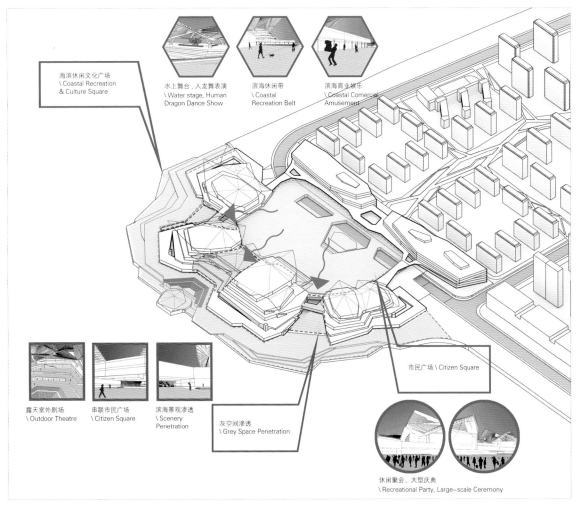

功能分区 \ Function Division

滨海水上剧场透视 \ Perspective of Coastal Theatre

文化艺术中心屋面观海透视 \ Perspective towards Sea on Cultural & Arts Center Roof

大剧院屋顶花园透视 \ Perspective of Roof Garden of Grand Theatre

多层级的露台、下沉庭院与伞状的屋顶相互穿插渗透，为滨海游客提供了层次丰富的遮阳避雨空间，合理的交通组织将不同标高的平台串联起来，使得游客可以全方位多层次欣赏海湾美景。

Multi-layer terrace, sunk courtyard and umbrella-shaped roof are interwoven and penetrated. Rational traffic organization connects activities in different platforms, in which visitors can enjoy view in different ways.

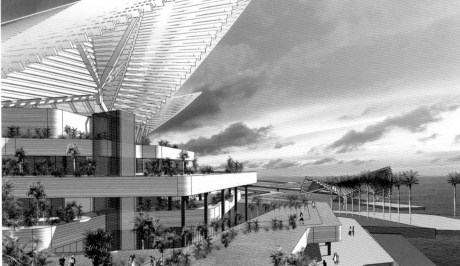

近景透视 \ Perspective of Close Shot

人点小透视 \ Detailed Perspective of Human Point

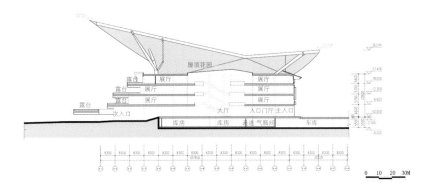

博物馆透视 \ Museum Perspective

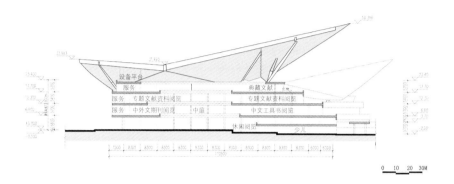

图书馆透视 \ Library Perspective

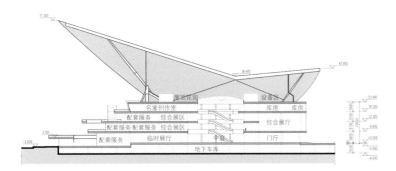

美术馆透视 \ Art Gallery Perspective

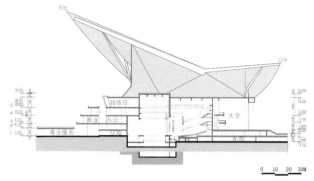

大剧院透视 \ Grand Theatre Perspective

作品年表（2009—2014）
CHRONOLOGICAL LIST OF PROJECTS (2009—2014)

2009
杭甬铁路客运专线概念方案
地点：杭甬线
合作者：殷建栋 刘鹏飞 单晓宇 程跃文 叶俊 胡晓明
建筑面积：18300m²
设计时间：2009 年
未实施

宁波市中级人民法院方案
地点：浙江宁波
合作者：陈立群 刘鹏飞 叶俊
　　　　胡 汉
建筑面积：57070m²
设计时间：2009 年
方案未中标

2010
杭州师范大学仓前校区 B 组团
地点：浙江杭州
合作者：王大鹏 柴敬 周炎鑫 庄允峰
　　　　谢悦 沈一凡 黄斌 王岳锋
　　　　陶涛 汤焱 祝容 胡晓明
　　　　贾秀颖 周霖
建筑面积：407916m²
设计时间：2010
施工中

北戴河铁路旅客站方案
地点：河北北戴河
合作者：殷建栋 叶俊
建筑面积：11970m²
设计时间：2009 年
未实施

湘潭城市规划馆及博物馆
地点：湖南湘潭
合作单位：湘潭市建筑设计院
合作者：王大鹏 柴敬 王禾苗
　　　　杨思思 胡晓明 叶俊
　　　　周霖 言海燕
建筑面积：40000m²
设计时间：2010
基本竣工

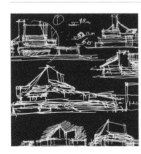

四川青川县博物馆方案
地点：四川省青川县
合作者：叶俊 胡晓明 单晓宇
建筑面积：4000m²
设计时间：2009 年
未实施

2011
中国港口博物馆
地点：浙江宁波
合作者：陈玲 刘翔华 叶俊
　　　　张朋君
建筑面积：40996m²
设计时间：2011
竣工时间：2014

西安大明宫遗址公园博物馆方案
地点：陕西西安
合作者：薄宏涛 刘鹏飞 唐斌
　　　　于晨 蒋珂 单晓宇
　　　　应瑛 杨涛 徐勤力
建筑面积：102000m²
设计时间：2009 年
未实施

黄岩博物馆
地点：浙江黄岩
合作者：陈玲 刘辉瑜 闵杰
　　　　王忠杰
建筑面积：10000m²
设计时间：2011
即将竣工

浙江建德城市规划展览馆及博物馆方案

地点：浙江建德
合作单位：浙江省建筑设计院
合作者：金 坤 陈 玲 叶 俊
　　　　胡蓓蓓 朱周胤
建筑面积：43000m²
设计时间：2011年
未实施

海南陵水夏宫酒店方案

地点：海南陵水
合作者：古振强 汤 焱 林嘉伟 余乐伟
建筑面积：46499m²
设计时间：2011年
未实施

苏步青纪念馆

地点：浙江平阳
合作者：陈 玲 刘辉瑜 王忠杰
　　　　朱文婧 张天钧 李 照
　　　　宋一鸣
建筑面积：4611m²
设计时间：2011
即将竣工

宁波东部新城 B1-4 地块方案

地点：浙江宁波
合作者：殷建栋 朱祯毅 黄 斌
　　　　古振强 李 照
建筑面积：72512m²
设计时间：2011年
未实施

山东临沂沂蒙革命历史纪念馆

地点：山东临沂
合作者：王大鹏 沈一凡 柴 敬
　　　　谢潘扬 殷建栋
建筑面积：43019m²
设计时间：2011
竣工时间：2014

2012

杭州市博物院方案

地点：浙江杭州
合作者：王大鹏 孟 浩 汤 焱
　　　　祝 容 郑克卿
建筑面积：155650m²
设计时间：2012年
未实施

山东莱州书法博物馆

地点：山东莱州
合作者：薄宏涛 于 晨 蒋 柯
　　　　樊文婷
建筑面积：17377m²
设计时间：2011年
未实施

福建龙岩展览城

地点：福建龙岩
合作者：薄宏涛 吴志全 李相鹏
　　　　祝 容 刘晶晶 王建国
建筑面积：417131m²
设计时间：2012年
未实施

湖南昭山两型发展中心
地点：湖南昭山
合作者：王大鹏 沈一凡 孟 浩
　　　　汤 焱 祝 容 裘 昉
建筑面积：52840m²
设计时间：2012 年
基本竣工

无锡锡东新城文化中心概念方案
地点：江苏无锡
合作者：殷建栋 朱文婧 郑克卿
　　　　董雍娴 周 慧 王碧君
建筑面积：41300m²
设计时间：2012 年
未实施

山西太原晋阳新城展示馆概念方案
地点：陕西太原
合作者：陈 玲 朱文婧 裘 昉
　　　　祝 容 汤 焱 言海燕
建筑面积：3000m²
设计时间：2012 年
未实施

南京河西文化艺术中心
地点：江苏南京
合作者：王大鹏 沈一凡 刘翔华
　　　　王 静 王忠杰 周 霖
　　　　汤 焱 祝 容 柴 敬
建筑面积：38800m²
设计时间：2012 年
竣工时间：2014 年

福建龙岩市委党校
地点：福建龙岩
合作单位：中建东北院厦门分院
合作者：陈 玲 黄斌毅 祝狄峰
　　　　朱祯毅 李 聪 陈凤婷
　　　　张朋君
建筑面积：98347m²
设计时间：2012
设计中

2013

南京陵园新村地块建筑方案
地点：江苏南京
合作者：王大鹏 沈一凡 刘翔华
　　　　汤 焱 祝 容 董雍娴
设计时间：2013 年
建筑面积：7700m²
未实施

厦门悦海湾酒店
地点：福建厦门
合作单位：中建东北院厦门分院
合作者：殷建栋 吴妮娜 杨 涛 庄允峰
　　　　闵 杰 周 逸 朱文婧 裘 昉
　　　　袁 越 陈 鑫 刘翔华 曾德鑫
　　　　郑建国 郑克卿 周 慧
建筑面积：90635m²
设计时间：2012
设计中

苏州越城遗址博物馆
地点：江苏苏州
合作者：殷建栋 钟承霞 朱文婧
　　　　桂汪洋 刘翔华
建筑面积：4736m²
设计时间：2013 年
设计中

湘潭市政务服务中心

地点：湖南湘潭
合作者：王大鹏 柴 敬 沈一凡
　　　　黄 斌 王岳峰
设计时间：2013年
建筑面积：103792m²
未实施

上海杉杉控股总部大楼

地点：上海
合作者：殷建栋 杨 涛 刘翔华
　　　　朱文婧 袁 越 陈 鑫
　　　　王 政 古振强 周 逸
建筑面积：105146m²
设计时间：2013
设计中

杭州钱江金融城概念方案

地点：浙江杭州
合作单位：北京土人城市规划设计有限公司
　　　　　艾奕康咨询（深圳）有限公司上海分公司
合作者：薄宏涛 王大鹏 郑英玉 杨 涛 刘翔华 柴 敬
　　　　黄 斌 王岳峰 李雯雯 戚卫娟 王 政 桂汪洋
　　　　梁超凡 汪 毅
建筑面积：1270000m²
设计时间：2013年
参加国际竞标入围

2014

湛江文化艺术中心方案

地点：广东湛江
合作者：王大鹏 杨 涛 刘翔华
　　　　沈一凡 孙 铭 吕思扬
　　　　刘鹤群
建筑面积：194084m²
设计时间：2014年
设计中

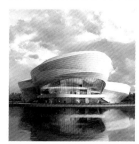

阜阳科技文化中心

地点：安徽阜阳
合作者：钟承霞 王大鹏 刘翔华
　　　　柴 敬 郭 晓 吕思扬
　　　　刘鹤群 黄卿云 张杰亮
建筑面积：50890m²
设计时间：2014年
设计中

长春国际雕塑博物馆

地点：吉林长春
合作者：王大鹏 吴旭斌 汪 毅
　　　　谢潘扬 吕思扬 黄卿云
　　　　张杰亮 刘鹤群
建筑面积：18858m²
设计时间：2014
设计中

南京栖霞广厅

地点：江苏南京
合作者：殷建栋 谢潘扬 郑国活
　　　　黄卿云 张杰亮 胡志菲
建筑面积：370000m²
设计时间：2014
设计中

南京永济江流景区概念规划及永济市复建方案设计

地点：江苏南京
合作者：王 静 蒋 楠 费移山
　　　　周 霖
建筑面积：134000 m²
设计时间：2014
设计中

后记
POSTSCRIPT

一．本书与中国建筑工业出版社出版的《中国建筑师——程泰宁》、《程泰宁建筑作品选 1997-2000》、《程泰宁建筑作品选 2001-2004》、《程泰宁建筑作品选 2005-2008》相衔接，并保持体例上的一致。

二．本书的版式设计由王琼宇、陈畅、赵伟伟等同志负责。插图分别由柴敬、骆晓怡、杨涛、刘翔华、张昊楠、王忠杰等同志绘制。

三．赖少艾、黄卿云同志为英文内容做了详细的校审。

四．中国建筑工业出版社对本书的出版给予了大力支持。

谨对以上同志表示诚挚的感谢。

I. This book links up and maintains consistent style with the *Chinese Architect : Cheng Taining, Cheng Taining Architecture Design Works in the Period of 1997-2000, Cheng Taining Architecture Design Works in the Period of 2001-2004 and Cheng Taining Architecture Design Works in the Period of 2005-2008* published by the China Architecture and Building Press.

II. The layout of this book was designed by Wang Qiongyu, Chen Chang, and Zhao Weiwei. The illustrations were drawn by Chai Jing, Luo Xiaoyi, Yang Tao, Liu Xianghua, Zhang Haonan, Wang Zhongjie and so on.

III. The English content in this book was proofread by Lai Shaoai and Huang Qingyun.

IV. The China Architecture and Building Press has given strong support for publication of this book.

We would like to express our sincere gratitude to all of them.

图书在版编目（CIP）数据

程泰宁建筑作品选2009-2014 / 程泰宁著. -- 北京：中国建筑工业出版社, 2015.5
ISBN 978-7-112-18039-4

Ⅰ.①程… Ⅱ.①程… Ⅲ.①建筑设计—作品集—中国—2009~2014 Ⅳ.①TU206

中国版本图书馆CIP数据核字(2015)第079871号

责任编辑：徐明怡 徐纺

程泰宁建筑作品选 2009-2014

程泰宁 著
*
中国建筑工业出版社出版、发行（北京西郊百万庄）
各地新华书店、建筑书店经销
上海雅昌艺术印刷有限公司 制版、印刷
*
开本：787×1092毫米 1/12 印张：24 字数：720千字
2015年12月第一版 2015年12月第一次印刷
定价：280.00元
ISBN 978-7-112-18039-4
(27276)
版权所有 翻印必究
如有印装质量问题，可寄本社退换
（邮政编码 100037）